MATH-
STAT.
LIBRARY

# Multiple Scattering Processes
## Inverse and Direct

MATH-
STAT.
LIBRARY

# APPLIED MATHEMATICS AND COMPUTATION

A Series of Graduate Textbooks, Monographs, Reference Works

Series Editor: ROBERT KALABA, University of Southern California

No. 1 MELVIN R. SCOTT
*Invariant Imbedding and its Applications to Ordinary Differential Equations: An Introduction, 1973*

No. 2 JOHN CASTI and ROBERT KALABA
*Imbedding Methods in Applied Mathematics, 1973*

No. 3 DONALD GREENSPAN
*Discrete Models, 1973*

No. 4 HARRIET H. KAGIWADA
*System Identification: Methods and Applications, 1974*

No. 5 V. K. MURTHY
*The General Point Process: Applications to Bioscience, Medical Research, and Engineering, 1974*

No. 6 HARRIET H. KAGIWADA and ROBERT KALABA
*Integral Equations Via Imbedding Methods, 1974*

No. 7 JULIUS T. TOU and RAFAEL C. GONZALEZ
*Pattern Recognition Principles, 1974*

No. 8 HARRIET H. KAGIWADA, ROBERT KALABA, and SUEO UENO
*Multiple Scattering Processes: Inverse and Direct, 1975*

No. 9 DONALD A. PIERRE and MICHAEL J. LOWE
*Mathematical Programming Via Augmented Lagrangians: An Introduction with Computer Programs, 1975*

*Other Numbers in preparation*

# Multiple Scattering Processes

## Processes

### Inverse and Direct

HARRIET H. KAGIWADA
The Rand Corporation
Santa Monica, California

ROBERT KALABA
University of Southern California
Los Angeles, California

SUEO UENO
Research Institute for Information Science
Kanazawa Institute of Technology
Ishikawa, Japan

 1975

**Addison-Wesley Publishing Company**
Advanced Book Program
Reading, Massachusetts

London · Amsterdam · Don Mills, Ontario · Sydney · Tokyo

MATH-
STAT.
LIBRARY

MATH-STAT

Library of Congress Cataloging in Publication Data

Kagiwada, Harriet H    1937-
    Multiple scattering processes.

    (Applied mathematics and computation ; no. 8)
    Bibliography:  p.
    1.  Inverse problems (Differential equations)
2.  Initial value problems.  3.  Scattering (Physics)
4.  Transport theory.  I.  Kalaba, Robert E., joint
author.  II.  Ueno, Sueo, 1911-     joint author.
III.  Title.
QA377.K33          515'.35              75-22363
ISBN 0-201-04104-9
ISBN 0-201-04105-7 pbk.

Reproduced by Addison-Wesley Publishing Company, Inc., Advanced Book Program, Reading, Massachusetts, from camera-ready copy prepared by the authors.

American Mathematical Society (MOS) Subject Classification Scheme (1970): 65L05, 65P05

Copyright © 1975 by Addison-Wesley Publishing Company, Inc.
Published simultaneously in Canada.

All rights reserved. No part of this publication may be reproduced, stored in a retrieval system, or transmitted, in any form or by any means, electronic, mechanical, photocopying, recording, or otherwise, without the prior written permission of the publisher, Addison-Wesley Publishing Company, Inc., Advanced Book Program, Reading, Massachusetts 01867, U.S.A.

Manufactured in the United States of America

QA377
K331
Math/
Stat

CONTENTS

v

1254

# SERIES EDITOR'S FOREWORD

Execution times of modern digital computers are measured in nano-seconds. They can solve hundreds of simultaneous ordinary differential equations with speed and accuracy. But what does this immense capability imply with regard to solving the scientific, engineering, economic, and social problems confronting mankind? Clearly, much effort has to be expended in finding answers to that question.

In some fields, it is not yet possible to write mathematical equations which accurately describe processes of interest. Here, the computer may be used simply to simulate a process and, perhaps, to observe the efficacy of different control processes. In others, a mathematical description may be available, but the equations are frequently difficult to solve numerically. In such cases, the difficulties may be faced squarely and possibly overcome; alternatively, formulations may be sought which are more compatible with the inherent capabilities of computers. Mathematics itself nourishes and is nourished by such developments.

Each order of magnitude increase in speed and memory size of computers requires a reexamination of computational techniques and an assessment of the new problems which may be brought within the realm of solution. Volumes in this series will provide indications of current thinking regarding problem formulations, mathematical analysis, and computational treatment.

ROBERT KALABA

PREFACE

We have written this book with three principal objectives in mind -

o   to formulate inverse problems in radiative trans-
    fer and to show the effectiveness of quasilineariza-
    tion in their computational resolution;

o   to introduce the functions  b  and  h  in book form
    and to indicate their fundamental role in the
    theory of multiple scattering;

o   to derive initial value problems to replace the
    more traditional boundary value problems and
    integral equations of multiple-scattering and to
    demonstrate their computational efficacy.

It turns out that the three objectives are intimately connected.
For, in order to treat inverse problems, it is important to
first reduce related direct problems to initial value problems,
and it is precisely in these problems that the functions  b
and  h  arise.

The function  b  has a physical meaning.  It is the
internal intensity in a homogeneous slab which is illuminated
uniformly at the top by isotropic sources of radiation.  It
is a function of altitude, thickness and a direction cosine.
Similarly, the function  h  is the internal intensity in a
slab which is uniformly illuminated by isotropic sources on
the bottom.  Due to the symmetry of the geometry, these two
functions are really one, but it is convenient to give them
separate names.  The problem of determining the  b  and  h
functions is designated the *basic problem*, the reasons for

xi

which will be made obvious.  These functions are of interest
in themselves.  But more importantly, *the functions  b  and
h  also solve the monodirectional illumination problem!*  In
terms of the functions  b  and  h,  the functions of interest
in the auxiliary problem of monodirectional illumination -
reflection, transmission, source,  X, Y, Φ  and internal intensity functions - are all simple algebraic functions of  b
and  h.  Consider the internal intensity function for the monodirectional illumination case, a function of altitude, thickness and two directions, the input and the propagation directions.  It is reduced to functions of fewer variables through
a unique new separation of variables formula, one which is
neither intuitive nor obvious.

The functions  b  and  h  will be encountered many
times in this book in discussions of different physical
problems.  Not only do they appear in the chapters on the
basic and auxiliary problems, but they are important in the
treatment of multiple scattering processes in media containing internal emitting sources and in homogeneous slabs bounded
by reflectors.  These functions play corresponding basic roles
in anisotropic scattering and other multiple scattering processes, as will be discussed in detail in a subsequent book.

The present theory is an outgrowth of invariant imbedding studies of a decade or two ago which are said to have
originated in the 1943 paper of V. Ambarzumian.  In the case
of monodirectional illumination of a homogeneous slab, he
showed that the diffusely reflected radiation may be determined without determining the entire radiation field within
the slab.  He did this by observing that the reflected radiation is unchanged if a layer is added to the top and an
identical layer is removed from the bottom.

We use a related procedure in this book. We add a
layer to one side of the slab,[*] enumerate the changes in the
physical quantities of interest, and then let the thickness
of the added layer tend to zero. This produces a differential equation, more frequently an integro-differential equation. Analysis of a slab of thickness zero provides an initial
condition for the initial value problem. Through approximation of integrals by sums via an appropriate quadrature formula, there results an initial value problem for a system of
ordinary differential equations which is readily solved computationally.

However, these physical manipulations can be avoided
or verified analytically starting with a traditional boundary
value problem or integral equation. The solution of the
initial value (Cauchy) problem can be proved to solve the
original problem, as in Chapter 7.

A great advantage in taking the Cauchy approach is
the concomitant ability to solve inverse problems. Inverse
problems for the estimation of sources, optical thickness,
anisotropic and other scattering parameters, based on measurements of multiply-scattered radiation, can be systematically
formulated and solved by use of modern system identification
techniques such as quasilinearization, which is described at
length.

The book is relatively self-contained, though the
reader may wish to consult V.V. Sobolev's A Treatise on Radiative Transfer (D. Van Nostrand Co., 1963) and a book on quadrature and numerical integration. There are exercises throughout
to serve as guides to working out some of the basic concepts
and problems to provide suggestions for computational tasks.

---

[*]Adding a thin slab at the top causes isotropic sources to
appear.

The first chapter treats multiple scattering processes in a one-dimensional medium. It introduces some of
the basic concepts of this book. Rates of production of
scattered particles, the reflection function, internal intensities, and other important quantities are regarded as
functions of the length of the medium. The initial value
problems and integral equations which they satisfy are
derived both physically and analytically. Then the relevant
Cauchy systems are obtained.

The auxiliary problem of isotropic scattering in
homogeneous slabs illuminated by parallel rays of radiation
is the subject of Chapter 2. The source function  J,  the
X  and  Y  functions, the internal intensities and the reflection and transmission functions all are shown to be solutions of initial value problems. The computational procedure
involves solving systems of ordinary differential equations,
subject to complete sets of initial conditions, by use of
high order numerical integration methods. Such systems result
from the application of Gaussian quadrature formulas. The
initial value method is extremely stable, and the computational
results are highly accurate. Numerous tables and graphs are
presented.

The basic problem of Chapter 3 refers to the theory of
the functions  b  and  h.  Physically, these are internal
intensities in homogeneous slabs which are illuminated by
isotropic sources of radiation either at the top or at the
bottom. Mathematically, these functions prove to be the
basic functions of the auxiliary problem of monodirectional
illumination. All of the functions introduced in Chapter 2
may be expressed *algebraically* in terms of  b  and  h.  The
b  and  h  functions are readily computed as solutions of

initial value problems, although they also satisfy a system
of integral equations.  Graphs of these functions are
presented in this Chapter, while the FORTRAN program for
computing them and extensive tables are to be found in
Appendices B and C.  These tables were produced with a seven-
point Gaussian quadrature formula, whose points, weights and
corresponding angles are listed in Appendix A.  These tables,
together with the algebraic formulas and a pocket calculator,
provide the functions of interest for the auxiliary problem.

  Chapter 4 is concerned with both inverse and direct
problems of multiple scattering in slabs containing internal
sources; again the ubiquitous functions  b  and  h  appear.
In the direct problem of known sources, the internal and external radiation fields are determined.  In an inverse
problem, the external (or internal) radiation fields are
given, and the task is to estimate the distribution of
sources.  This is formulated as a nonlinear least squares
boundary value problem, which is then solved by the method
of quasilinearization.  Numerical results show the effectiveness of the procedure.  Tables of the resolvent kernel are
also presented in Chapter 4.

  Multiple scattering in inhomogeneous media is considered in Chapter 5.  Of special interest are the inverse
problems for the estimation of layers and total thickness
of inhomogeneous slabs.  Multiple scattering processes with
Lambert's law and specular reflectors underlying slabs are
treated in Chapter 6.  In the former case, the theory permits
a certain reduction involving functions of fewer variables.
In the case of conservative specular reflectors, solutions
can be expressed simply in terms of the functions  b  and  h
of the basic problem; similarly for the Lambert law case.

Anisotropic scattering is the subject of Chapter 7. Here, the functions of physical interest depend on several angular arguments. Two of the ways of reducing the number of arguments are consideration of axially symmetric fields and expansion in Legendre functions. The functions b and h play a major role in anisotropic scattering, a point which will be described in detail in a subsequent book.

Gaussian quadrature data for a seven point formula are given in Appendix A. Appendix B describes and lists the FORTRAN program for computing the functions b and h. Tables of the functions b and h are presented in Appendix C.

We are indebted to A. Fymat for a careful reading of the manuscript, to L. Anderson for the typing, and to our colleagues who have directly or indirectly contributed to our researches in multiple scattering.

# CHAPTER 1

## INTRODUCTION: ONE-DIMENSIONAL TRANSPORT THEORY

## 1.1    ONE-DIMENSIONAL TRANSPORT

We shall first discuss a one-dimensional transport process in detail, deriving Cauchy problems for various functions of interest, and taking up certain computational aspects. Consider a homogeneous rod extending from 0 to L. Assume that when a particle moves through a distance $\Delta$ along the rod, there is a probability $\Delta + o(\Delta)$ that it will interact with the rod. The function $o(\Delta)$ , of course, is such that the ratio $o(\Delta)/\Delta$ tends to zero as $\Delta$ tends to zero. The result of an interaction is that the particle is removed from the process. Then, with probability $\lambda$, $0 \leq \lambda \leq 1$, it is re-emitted (scattered), and it is equally likely to go to the right or to the left (i.e., the scattering is isotropic). Particles which emerge from the right or left end of the rod do not re-enter the rod. Along the rod there is a primary distribution of sources. We let

$g(t)dt$ = the primary rate of production of particles
going to the right in the spatial interval
$(t,t+dt)$,  $0 \leq t \leq L$. The sources are isotropic,
so that the same rate of particles going to
the left is produced.                         (1.1)

Alternatively we may state that $g(t)$ is the number of particles produced at $t$ and going to the right, per unit length
per unit time.  Our task is to investigate the steady state
multiple-scattering process which ensues.  We wish to determine the fields both within and without the rod.

In the analysis to follow, the length of the rod is
regarded as an independent variable, $x$,  and  $0 \leq x \leq L$.  Let us
introduce the source function  $u(t,x)$,  $0 \leq t \leq x$,  by means of
the definition

$u(t,x)$ $\Delta$ = the total rate of production of scattered
particles going to the right in the inter-
val  $(t,t+\Delta)$  in a rod of length  $x$  and
due to the distribution of primary sources
$g(t)$,  $0 \leq t \leq x$ .                     (1.2)

The parameter  $t$,  once chosen, is held fixed in the course
of the analysis.  We wish to study the effect on  $u(t,x)$  of
varying the length  $x$.  We conceptually increase the length
of the rod from  $x$  to  $x + \Delta$.  The rate of production of
scattered radiation at the point  $t$  is greater for the longer
rod, due to the particles which are produced in the segment
of rod between  $x$  and  $x + \Delta$  with the rate  $u(x,x)\Delta + o(\Delta)$.
These particles act as sources which give rise to additional
scattering at the point  $t$.

We introduce another source function $J(t,x)$:

$J(t,x) \Delta$ = the rate of production of particles in the
interval $(t,t+\Delta)$ going to the right due
to one incident particle per unit time at
the right end of the rod, $0 \le t \le x$.    (1.3)

In terms of this source function, the contribution to $u(t,x+\Delta)$
due to the primary and secondary sources in the rod segment
between $x$ and $x + \Delta$ is just $u(x,x) \Delta J(t,x) + o(\Delta)$.
Clearly, the rate of production per unit length at $t$ in
the rod of length $x + \Delta$ can be expressed in the form

$$u(t,x+\Delta) = u(t,x) + u(x,x) \Delta J(t,x) + o(\Delta).    (1.4)$$

Expand the left hand side of the above equation

$$u(t,x+\Delta) = u(t,x) + \Delta u_x(t,x) + \frac{1}{2} \Delta^2 u_{xx}(t,x) + \ldots$$

$$= u(t,x) + \Delta u_x(t,x) + o(\Delta) ,    (1.5)$$

where subscripts indicate partial differentiation. Rewrite
Eq. (1.4), after dropping $u(t,x)$ from both sides, as

$$\Delta u_x(t,x) = \Delta u(x,x) J(t,x) + o(\Delta).    (1.6)$$

Divide through by $\Delta$, and let $\Delta$ tend to zero. The result
is a basic differential equation of the theory,

$$u_x(t,x) = u(x,x) J(t,x) ,    \qquad x \ge t .    (1.7)$$

We must now investigate  u(x,x)  and  J(t,x)  further as
functions of  x.

Let us first consider the function  J(t,x).  Increase
the length of the rod from  x  to  x + Δ,  and keep the same
unit source acting on the right end.  Conceptually, we have
the two cases of a unit source incident on a rod of length
x,  and a unit source incident on a rod of length  x + Δ,
whose physical properties on the interval  t = 0  through
t = x  are identical to those of the rod of length  x.  We
can write a relation linking source functions for these rods.
It is

$$J(t,x+\Delta) = (1-\Delta)J(t,x) + J(x,x)\ \Delta J(t,x) + o(\Delta),$$

$$t<x \ . \quad (1.8)$$

The first term on the right hand side occurs because the
unit source at  t = x + Δ  is reduced to (1 - Δ)  as it passes
through the rod segment between  t = x + Δ  and  t = x.  We
regard the rate of scattering at  t  to be due in part to
this reduced source acting on the rod on length  x  and producing the scattered particles at the internal point  t.  An
additional contribution is due to the source in (x, x+Δ) as
shown in the second term.  Other contributions are of orders
$\Delta^2$,  $\Delta^3$,  ... and are included in the term  o(Δ).  A limiting
form of Eq. (1.8) is

$$J_x(t,x) = -J(t,x) + J(x,x)\ J(t,x) \ , \qquad t<x \ . \quad (1.9)$$

We next consider the function  J(x,x) as a function
of  x.  For this purpose we introduce the reflection function
r ,

$r(x)$ = the average rate of particles reflected from the rod of length $x$, due to one incident particle per unit time at the right end.

(1.10)

The function $J(x,x)$ may be expressed in terms of $r(x)$ by means of the relation

$$J(x,x) = \frac{\lambda}{2} + \frac{\lambda}{2} r(x) \ .$$

(1.11)

This follows by consideration of the total rate of production in a segment of length $\Delta$ at the right end of the rod. On the one hand it is $J(x,x) \Delta$. On the other hand it is, to first order in $\Delta$, the sum of the products $1 \cdot \Delta \cdot \lambda \cdot \frac{1}{2}$ and $r(x) \cdot \Delta \cdot \lambda \cdot \frac{1}{2}$. The first product is due to the particle incident at the right end of the rod and having a probability $\Delta$ of being absorbed, a probability $\lambda$ of being reemitted, and a probability $1/2$ of then going either to the right or to the left. Similarly, the second product is due to the particles going to the right at the right end; they may interact and then be scattered there.

The function $r(x)$ satisfies the functional relation

$$r(x+\Delta) = (1-\Delta)\, r(x) \left\{ (1-\Delta) + \Delta \left[ \frac{\lambda}{2} + \frac{\lambda}{2} r(x) \right] \right\}$$
$$+ \Delta \left[ \frac{\lambda}{2} + \frac{\lambda}{2} r(x) \right] + o(\Delta) \ .$$

(1.12)

This follows from considering that the incident particle first does not interact in the interval of length $\Delta$ and then that it does. In the event that the incident particle does not interact in the interval $(x,x+\Delta)$, it produces reflected particles which may or may not interact in $(x, x+\Delta)$, it may be directly back-scattered, or it may produce other reflected particles. It follows from Eq. (1.12) that the function $r(x)$ satisfies the Riccati equation

$$\frac{dr}{dx} = \frac{\lambda}{2} + (\lambda - 2) \ r + \frac{\lambda}{2} \ r^2 , \qquad\qquad 0 < x . \quad (1.13)$$

The initial condition which $r$ satisfies is clearly

$$r(0) = 0 , \qquad\qquad\qquad (1.14)$$

which is due to the fact that no particles are reflected from a rod of length zero. Equations (1.13) and (1.14) form an initial value problem which is readily resolved by modern analog and digital computers. The solution at $x = L$, $r(L)$, is the reflection function for the desired rod of length L. While it is possible to obtain the analytical solution for the above reflection function, it may be quite difficult to do so in the multidimensional case.


1.2    NUMERICAL SOLUTION

To illustrate the basic concept for the numerical solu-tion of a system of ordinary differential equations subject to initial conditions, we discuss the classical method of Euler in conjunction with the above equation for $r(x)$. We choose a step length h, sufficiently small for our purposes. We are given that $r(0) = 0$. The approximate value of $r$ when $x = h$ is given by the formula

$$r(h) = r(0) + h \left. \frac{dr}{dx} \right|_{x=0} , \qquad\qquad (1.15)$$

or

$$r(h) \cong r(0) + h\left[\frac{\lambda}{2} + (\lambda-2)r(0) + \frac{\lambda}{2} r^2(0)\right] = \frac{\lambda}{2} h ,$$

$$(1.16)$$

where we have deleted the approximation sign.  We may proceed
to calculate  $r(2h)$, ..., by use of the recurrence formula

$$r((n+1)h) = r(nh) + h\left[\frac{\lambda}{2} + (\lambda-2)r(nh) + \frac{\lambda}{2} r^2(nh)\right],$$

$$(1.17)$$

for  $n = 1,2,...,$ .  Methods of great accuracy, stability and
efficiency are available for the numerical integration of
ordinary differential equations.  The initial value problems
of this book may be considered solvable by one of these methods.


1.3    DERIVATION CONTINUED

Returning to the source function  $J(t,x)$,  t fixed and
$t \leq x$ ,  we note that it could be determined as the solution of
the initial value problem

$$J_x(t,x) = - J(t,x) + \left[\frac{\lambda}{2} + \frac{\lambda}{2} r(x)\right] J(t,x) ,$$

$$t < x , \quad (1.18)$$

$$J(t,t) = \frac{\lambda}{2} + \frac{\lambda}{2} r(t) .$$

$$(1.19)$$

The initial condition of Eq. (1.19) follows from putting
x = t  in Eq. (1.11), the relation between  $J(x,x)$   and
$r(x)$.  Regard  $r(x)$  as a known function of  x   for

$0 \leq x \leq L$. Its value at $x = t$ gives $J(t,t)$. Solution of the differential Eq. (1.18) yields $J(t,x)$ for various $x$, $t \leq x \leq L$.

The calculation of the functions $r$ and $J$ could also be done simultaneously, thereby eliminating the problem of storing functional values of $r$. The procedure is then as follows. The differential equation for $r$ is integrated from $x = 0$ to $x = t$, where $t$ is a previously chosen value. The value of $J(t,t)$ is determined from Eq. (1.19). Then integrate the system of two equations for $r(x)$ and $J(t,x)$ from $x = t$ to $x = L$. If evaluation of the source function is desired at another internal point $t'$, $t < t' < x$, interrupt the calculation at $x = t'$, evaluate $J(t',t')$, adjoin the differential equation for $J(t',x)$ to the existing system and continue the integration until $x = L$. Thus the determination of $J(t,x)$ could be made for a selected set of values of $t$, as well as $x$.

To determine the function $u(t,x)$, $x \geq t$, we return to Eq. (1.7). We have already discussed the second factor on the right side, $J(t,x)$. Let us now consider the first factor, $u(x,x)$.

Introduce the emission function $e$ by means of the definition

> $e(x)$ = the average number of particles per unit time emerging from the right end of the rod of length $x$ due to the sources $g(t)$, $0 \leq t \leq x$.
>
> (1.20)

Through a now familiar process we may write

$$u(x,x)\Delta = \Delta g(x) + e(x)\Delta \frac{\lambda}{2} + o(\Delta) . \qquad (1.21)$$

This leads to the representation formula

$$u(x,x) = g(x) + (\lambda/2)\, e(x) \ . \tag{1.22}$$

By extending the length of the rod from $x$ to $x + \Delta$ we find that

$$e(x + \Delta) = e(x) - \Delta\, e(x) + u(x,x)\Delta\, [1 + r(x)] + o(\Delta). \tag{1.23}$$

This leads to the desired differential equation for the function $e$,

$$e_x(x) = -e(x) + \left[g(x) + \frac{\lambda}{2} e(x)\right]\left[1 + r(x)\right], \qquad x \ge 0 \ . \tag{1.24}$$

The initial condition on the function $e$ at $x = 0$ is

$$e(0) = 0 \ . \tag{1.25}$$

The prescription for determining the function $u(t,x)$ for $t$ being a fixed positive number and all $x \ge t$ is this. On the interval $0 \le x \le t$ the differential equations (1.24) and (1.13) for the functions $r$ and $e$ are integrated subject to the initial conditions in Eqs. (1.25) and (1.14). In particular, the values of $e(t)$ and $r(t)$ are determined numerically. At $x = t$ to the differential equations for the functions $r$ and $e$ we adjoin the differential equations for the functions $J(t,x)$ and $u(t,x)$,

$$J_x(t,x) = -J(t,x) + \frac{\lambda}{2}\left[1 + r(x)\right] J(t,x) \tag{1.26}$$

and

$$u_x(t,x) = g(x) + \frac{\lambda}{2} e(x) J(t,x) .$$  (1.27)

The initial conditions on the functions $J$ and $u$ at $x = t$ are

$$J(t,t) = \frac{\lambda}{2} + \frac{\lambda}{2} r(t)$$  (1.28)

and

$$u(t,t) = g(t) + \frac{\lambda}{2} e(t) .$$  (1.29)

The right sides of the last two equations are known numerically. The integration of the system of four simultaneous ordinary differential equations continues until $x = L$.


1.4    ANALYTICAL REDERIVATION

The intuitive considerations of the previous pages can be put on a firm foundation by starting with a basic integral equation for the function $u(t,x)$. Let us now see how this may be done. By considering the total rate of production of particles moving to the right in the interval $(t,t+\Delta)$ it is seen that

$$u(t,x) \Delta = g(t) \Delta + \frac{\lambda}{2} \Delta \int_0^x e^{-|t-y|} u(y,x) \, dy + o(\Delta) .$$  (1.30)

The first term on the right hand side of Eq. (1.30) is obviously the contribution due to the primary sources. The

integral term is the mathematical expression for the contribution of the particles produced in the interval (y,y+dy)
going directly to the interval (t,t+Δ) and being scattered
there. To understand why the exponential factor is present,
consider the absorption of particles in a segment of length
ds. If N is the number of particles entering the segment,
then the change in N as they pass through the segment is
dN; and this change is given by the equation

$$- \frac{dN}{N} = ds \; ,$$

(1.31)

according to the described property of the rod. Over a finite
segment of length $s_1$, let the original and final numbers of
particles be $N_0$ and $N_1$ respectively. We have

$$- \int_{N_0}^{N_1} \frac{dN}{N} = \int_0^{s_1} ds \; ,$$

(1.32)

$$- \ln \frac{N_1}{N_0} = s_1 \; ,$$

(1.33)

$$N_1 = N_0 \, e^{-s_1} \; .$$

(1.34)

The last equation shows that the number of particles which
pass through a segment of length $s_1$ without absorption is
the original number multiplied by the fraction $e^{-s_1}$. Thus
the integral term of Eq. (1.30) may be regarded as arising
as follows (read the factors from right to left):

$$\int_0^x \frac{\lambda}{2} \cdot \Delta \cdot e^{-|t-y|} \cdot u(y,x) \; dy \; .$$

(1.35)

The limiting form of Eq. (1.30) is the Fredholm integral equation for the source function $u$,

$$u(t,x) = g(t) + \frac{\lambda}{2} \int_0^x e^{-|t-y|} u(y,x)\, dy,$$
$$0 \le t \le x. \quad (1.36)$$

Differentiate with respect to $x$ on both sides of the above equation to obtain the equation

$$u_x(t,x) = \frac{\lambda}{2} e^{-(x-t)} u(x,x) + \frac{\lambda}{2} \int_0^x e^{-|t-y|} u_x(y,x)\, dy,$$
$$(1.37)$$

where use has been made of Leibniz's theorem for the differentiation of an integral,

$$\frac{d}{dx} \int_{a(x)}^{b(x)} f(y,x)\, dy = f(b,x) \frac{db}{dx} - f(a,x) \frac{da}{dx}$$
$$+ \int_{a(x)}^{b(x)} f_x(y,x)\, dy. \quad (1.38)$$

We pause here to recall that linear Fredholm integral equations possess the following fundamental property: Let $u$, $u_1$ and $u_2$ be the solutions of the linear integral equations

$$u(t,x) = h(t) + \int_0^x k(t,y)\, u(y,x)\, dy, \quad (1.39)$$

$$u_1(t,x) = h_1(t) + \int_0^x k(t,y)\, u_1(y,x)\, dy, \quad (1.40)$$

$$u_2(t,x) = h_2(t) + \int_0^x k(t,y)\, u_2(y,x)\, dy. \quad (1.41)$$

If the forcing function $h(t)$ is a linear combination of $h_1(t)$ and $h_2(t)$,

$$h(t) = c_1 u_1(t) + c_2 h_2(t) , \qquad (1.42)$$

then the corresponding solution $u$ may be represented as the linear combination

$$u(t,x) = c_1 u_1(t,x) + c_2 u_2(t,x) . \qquad (1.43)$$

We will apply this theorem to Eq. (1.37), which we regard as an integral equation for the function $u_x$.

Introduce the new function $J$ as the solution of the integral equation

$$J(t,x) = \frac{\lambda}{2} e^{-(x-t)} + \frac{\lambda}{2} \int_0^x e^{-|t-y|} J(y,x) \, dy ,$$
$$0 \leq t \leq x . \quad (1.44)$$

Note that the forcing function is essentially the same as that in Eq. (1.37). There are physical interpretations of all of these analytical forms. In the integral equation for $J$, we regard primary sources as arising from the incident particles which travel directly to $t$ and are first scattered there. View Eq. (1.37) as an integral equation for the function $u_x(t,x)$. The solution expressed in terms of $J$ is, from the superposition principle,

$$u_x(t,x) = u(x,x) \, J(t,x) , \qquad x \geq t . \quad (1.45)$$

This is one of the basic equations which we obtained earlier by the addition of a segment of rod (Eq. 1.7).

In order to complete the initial value problem, we differentiate both sides of Eq. (1.44) to obtain the relation

$$J_x(t,x) = -\frac{\lambda}{2} e^{-(x-t)} + \frac{\lambda}{2} e^{-(x-t)} J(x,x)$$

$$+ \frac{\lambda}{2} \int_0^x e^{-|t-y|} J_x(y,x) \, dy \ . \tag{1.46}$$

Note the forcing functions and again use the superposition theorem. The solution of this integral equation for the function $J_x$ is expressed in terms of $J(t,x)$ as

$$J_x(t,x) = -J(t,x) + J(x,x) \, J(t,x) \ , \tag{1.47}$$

which is a differential equation for $J(t,x)$. To determine $J(x,x)$ we note that the integral Eq. (1.44) gives

$$J(x,x) = \frac{\lambda}{2} + \frac{\lambda}{2} \int_0^x e^{-(x-y)} J(y,x) \, dy \ . \tag{1.48}$$

Introduce the function $r(x)$ by the equation

$$r(x) = \int_0^x e^{-(x-y)} J(y,x) \, dy \ , \qquad\qquad x>0 \ , \tag{1.49}$$

so that Eq. (1.48) becomes a formula for $J(x,x)$ in terms of $r(x)$ ,

$$J(x,x) = \frac{\lambda}{2} \left[ 1 + r(x) \right] \ . \tag{1.50}$$

In Eq. (1.49), the number of reflected particles is expressed in terms of the number produced at a general internal point y which travel to the right end without further interaction. By differentiating Eq. (1.49) it is seen that

$$r'(x) = J(x,x) - \int_0^x e^{-(x-y)} J(y,x) \, dy$$

$$+ \int_0^x e^{-(x-y)} J_x(y,x) \, dy \ . \qquad (1.51)$$

Making use of Eqs. (1.49) and (1.47), this equation becomes

$$r'(x) = J(x,x) - r(x)$$

$$+ \int_0^x e^{-(x-y)} \left[ -J(y,x) + J(x,x) \, J(y,x) \right] \, dy$$

$$= J(x,x) - r(x) - r(x) + J(x,x) \, r(x) \ . \qquad (1.52)$$

Finally, in view of Eq. (1.50), the last equation becomes an ordinary differential equation for the function $r$,

$$r'(x) = \frac{\lambda}{2} \left[ 1 + r(x) \right]^2 - 2 \, r(x) \ , \qquad x>0 \ , \qquad (1.53)$$

which agrees with the differential equation for the function $r$ obtained earlier in Eq. (1.13). The initial condition at $x = 0$ is

$$r(0) = 0 \ , \qquad (1.54)$$

from Eq. (1.49). Notice that Eqs. (1.53) and (1.54) are an initial value problem for $r(x)$, a problem easily solved by modern digital and analog computers, as we saw earlier.

Now we turn to the second factor in the right-hand side of Eq. (1.45), $u(x,x)$. Write

$$u(x,x) = g(x) + \frac{\lambda}{2} \int_0^x e^{-(x-y)} u(y,x) \, dy$$

$$= g(x) + \frac{\lambda}{2} e(x) \, , \qquad (1.55)$$

where $e(x)$ is introduced as

$$e(x) = \int_0^x e^{-(x-y)} u(y,x) \, dy \, . \qquad (1.56)$$

Differentiation yields

$$e'(x) = u(x,x) - \int_0^x e^{-(x-y)} u(y,x) \, dy$$

$$+ \int_0^x e^{-(x-y)} u_x(y,x) \, dy \qquad (1.57)$$

In view of Eqs. (1.45), (1.56), and (1.49) this becomes

$$e'(x) = u(x,x) - e(x) + u(x,x) \int_0^x e^{-(x-y)} J(y,x) \, dy$$

$$= u(x,x) - e(x) + u(x,x) \, r(x) \, . \qquad (1.58)$$

Finally, using Eq. (1.55), we obtain the relation

$$e'(x) = - e(x) + \left[ g(x) + \frac{\lambda}{2} e(x) \right] [1 + r(x)] \, . \qquad (1.59)$$

Using Eq. (1.55), Eq. (1.45) is rewritten

$$,x) = J(t,x)\left[ g(x) + \frac{\lambda}{2} e(x)\right], \qquad t<x . \qquad (1.60)$$

All the needed differential equations and auxiliary relations are now at hand. Let us recapitulate.

The differential equations and initial conditions for $r(x)$ and $e(x)$ are used to determine those functions for $x$ beginning at zero, and on up to $x = t$ ,

$$r'(x) = \frac{\lambda}{2} [1 + r]^2 - 2r, \qquad r(0) = 0 , \qquad (1.61)$$

$$e'(x) = - e + \left[ g + \frac{\lambda}{2} e\right]\left[ 1 + r\right], \; e(0) = 0 . \qquad (1.62)$$

In particular, at this stage of the calculation the values of $r(t)$ and $e(t)$ are known numerically. At $x = t$ we adjoin the two differential equations (1.47) and (1.60) for $J(t,x)$ and $u(t,x)$ to those for $r$ and $e$ ,

$$J_x(t,x) = - J(t,x) + J(t,x) \frac{\lambda}{2} [1 + r] , \qquad (1.63)$$

$$u_x(t,x) = J(t,x) \left[ g(x) + \frac{\lambda}{2} e(x)\right] . \qquad (1.64)$$

For initial conditions on the functions $J$ and $u$ at $x = t$ we have

$$J(t,t) = \frac{\lambda}{2} [1 + r(t)] = \text{known} , \qquad (1.65)$$

$$u(t,t) = g(t) + \frac{\lambda}{2} e(t) = \text{known} . \qquad (1.66)$$

The integration of the four ordinary differential equations (1.61), (1.62), (1.63), and (1.64) continues until $x = L$. In this way the desired value of $u(t,L)$ is obtained, of course, much more. Not only are $J(t,L)$, $r(L)$ a and, obtained, but also $u(t,x)$, $J(t,x)$, $r(x)$ and $e(x)$.


## 1.5    SEVERAL ADDITIONAL FUNCTIONS:  X  AND  Y

In addition to the functions discussed earlier, several others play a significant role in the theory. First let us take up the  X  and  Y  functions, which are defined by the relations

$$X(x) = J(x,x) \tag{1.67}$$

$$Y(x) = J(0,x) . \tag{1.68}$$

As earlier,  $J(t,x)$  is the solution of the Fredholm integral equation

$$J(t,x) = \frac{\lambda}{2} e^{-(x-t)} + \frac{\lambda}{2} \int_0^x e^{-|t-y|} u(y,x) \, dy . \tag{1.69}$$

Thus  X  and  Y  are rates of production of scattered particles at the right and left ends of the rod, respectively. On physical grounds, by the addition of a segment at the right end of the rod, we see that

$$Y(x + \Delta) = Y(x) - Y(x) \, \Delta + X(x) \, \Delta Y(x) + o(\Delta) . \tag{1.70}$$

It follows that

(1.71)

$$\frac{dY}{dx} = -Y + XY \ , \qquad Y(0) = \frac{\lambda}{2} \quad .$$

...ction $x$ satisfies the equation

(1.72)

$$X(x + \Delta) = x(x) + Y(x) \ \Delta Y(x) + o(\Delta) \ ,$$

which follows from the addition of a segment to the left end.

This becomes

(1.73)

$$\overline{dx} = Y^2(x) \ , \qquad X(0) = \frac{\lambda}{2} \ .$$

Equations (1.71) and (1.73) are readily integrated numerically. Equation (1.71) follows from Eq. (1.44) with $t = 0$. To derive Eq. (1.73) we put $t = x - z$ in Eq. (1.69) above. This yields

$$J(x-z,x) = \frac{\lambda}{2} e^{-z} + \frac{\lambda}{2} \int_0^x e^{-|x-z-y|} J(y,x) \ dy \ . \quad (1.74)$$

In this integral we make the substitution

$$y = x - y' \ , \qquad\qquad\qquad (1.75)$$

which provides

$$J(x-z,x) + \frac{\lambda}{2} e^{-z} + \frac{\lambda}{2} \int_0^x e^{-|y'-z|} J(x-y',x) \ dy' \ .$$
$$(1.76)$$

By differentiating with respect to $x$ we obtain

$$\frac{d}{dx} J(x-z,x) = \frac{\lambda}{2} e^{-(x-z)} J(0,x)$$

$$+ \frac{\lambda}{2} \int_0^x e^{-|y'-z|} \frac{d}{dx} J(x-y',x) \, dy' \, .$$

$$(1.77)$$

The solution of this integral equation is

$$\frac{d}{dx} J(x-z,x) = J(0,x) \, J(z,x) \, .$$

$$(1.78)$$

By putting $z = 0$ we obtain the desired result

$$\frac{d}{dx} J(x,x) = J^2(0,x) \, .$$

$$(1.79)$$

The initial conditions at $x = 0$ follow directly from the integral equation (1.69).

The complete initial value problem for determining $u(t,x)$ can be written as

$$Y'(x) = -Y + XY \, , \qquad\qquad Y(0) = \frac{\lambda}{2} \, , \quad (1.80)$$

$$X'(x) = Y^2 \, , \qquad\qquad X(0) = \frac{\lambda}{2} \, , \quad (1.81)$$

$$e'(x) = -e + \left[ g + \frac{\lambda}{2} e \right] \frac{2}{\lambda} X \, , \qquad e(0) = 0 \, , \quad (1.82)$$

$$J_x(t,x) = -J + JX \, , \qquad\qquad J(t,t) = X(t) \, , \quad (1.83)$$

$$u_x(t,x) = J \left[ g + \frac{\lambda}{2} e \right], \qquad u(t,t) = g(t) + \frac{\lambda}{2} e(t) \, .$$

$$(1.84)$$

## 1.6    INTERNAL INTENSITY FUNCTIONS

Let us define the internal intensity functions

$m(t,L)$ = the average number of particles passing
the point  $t$  which are traveling to the
right in a unit of time, in a rod of length
$L$ , due to one incident particle at the
right end, per unit time.                (1.85)

$n(t,L)$ = the average number of particles passing the
point  $t$  which are traveling to the left in
a unit of time, in a rod of length  $L$ , due
to one incident particle at the right end,
per unit time,  $0 \leq t \leq L$ .                (1.86)

By considering the change in intensity as the particles
traveling to the right pass from the point  $t$  to  $t + \Delta$,
we may write the relation (suppressing the length L)

$$m(t+\Delta) = m(t) - \Delta m(t) + \Delta J(t) + o(\Delta) .        (1.87)$$

As before,  $J$  is the source function for the unit input.
The effects represented in Eq. (1.87) are the absorption in
the segment of length $\Delta$ , scattering in the same segment,
and higher order processes.  In the limit as  $\Delta$  tends to 0,
Eq. (1.87) yields

$$\dot{m} = - m + J ,                (1.88)$$

where the dot signifies differentiation with respect to  $t$ .
For the particles traveling to the left, we have

$$n(t) = n(t+\Delta) - \Delta n(t+\Delta) + \Delta J(t) + o(\Delta)$$

$$= n(t+\Delta) - \Delta n(t) + \Delta J(t) + o(\Delta) \ , \qquad (1.89)$$

$$-\dot{n} = -n + J \ . \qquad (1.90)$$

The boundary conditions, which describe the incident particles, are

$$m(0,L) = 0 \ , \qquad n(L,L) = 1 \ . \qquad (1.91)$$

Equations (1.88), (1.90) and (1.91) form a two-point boundary value problem whose analytical solution is

$$m(t,L) = \int_0^t J(y,L) \ e^{-(t-y)} \ dy \ , \qquad (1.92)$$

$$n(t,L) = \int_t^L J(y,L) \ e^{-(y-t)} \ dy + e^{-(L-t)} \ . \qquad (1.93)$$

We may regard Eqs. (1.92) and (1.93) as formulas which determine m and n once the function J is known. This may be considered to be a classical approach.

Another related approach is to rid the differential equations of J completely by noting that

$$J(t,L) = \frac{\lambda}{2} \left[ m(t,L) + n(t,L) \right] \ , \qquad (1.94)$$

which is clearly true, both from the physical and analytical aspects. Then the boundary value problem becomes the classical transport equations

$$\dot{m} = - m + \frac{\lambda}{2} (m + n) \ , \qquad m(0,L) = 0 \ , \qquad (1.95)$$

$$- \dot{n} = - n + \frac{\lambda}{2} (m + n) \ , \qquad n(L,L) = 1 \ . \qquad (1.96)$$

We could ask for an analytical or computational solution of
Eqs. (1.95) and (1.96). However, we wish to go on now to
derive new equations for the functions $m$ and $n$ by the
method of invariant imbedding.

Let us first derive an initial value problem by an
analysis of the physical process. Later, we shall give a
re-derivation starting with Eqs. (1.95) and (1.96). At
this time we pretend to know nothing of Eqs. (1.87) - (1.96).
Let the length of the rod be the variable $x$, $0 \leq x \leq L$. Increase the length to $x + \Delta$ and study the changes produced
in the internal intensity functions. The relations are

$$m(t,x+\Delta) = m(t,x) - \Delta m(t,x) + \Delta \frac{\lambda}{2} \left[ 1 + r(x) \right] m(t,x)$$

$$+ o(\Delta) \ , \qquad (1.97)$$

$$n(t,x+\Delta) = n(t,x) - \Delta n(t,x) + \Delta \frac{\lambda}{2} \left[ 1 + r(x) \right] n(t,x)$$

$$+ o(\Delta) \ , \qquad (1.98)$$

where $r(x)$ is the reflection function, as before. The
terms on the right hand sides show the effect of absorption
and scattering in the additional segment between $(x, x + \Delta)$.
The scattering is due to the interactions of the incident
particles in $(x, x+\Delta)$. These interactions produce
$\Delta(\lambda/2)(r+1)$ particles going to the left, which may be considered to be incident on the rod of length $x$. Their effect
is given by the third term on the right in Eqs. (1.97) and (1.98).

Passage to the limit leads to the ordinary differential equations (t is held fixed)

$$m_x(t,x) = - m(t,x) + \frac{\lambda}{2} \left[1 + r(x)\right] m(t,x) \ , \qquad (1.99)$$

$$n_x(t,x) = - n(t,x) + \frac{\lambda}{2} \left[1 + r(x)\right] n(t,x) \ ,$$
$$x \geq t \ . \qquad (1.100)$$

The initial conditions, when the length $x = t$ , are

$$m(t,t) = r(t) \ , \qquad\qquad\qquad\qquad (1.101)$$

$$n(t,t) = 1 \ , \qquad\qquad\qquad\qquad (1.102)$$

the former through the definitions of $m$ and $r$ , the latter through the description of the source. To form the complete initial value problem, we must adjoin the equations for $r$ from an earlier section,

$$r'(x) = \frac{\lambda}{2} \left[1 + r\right]^2 - 2r \ , \qquad\qquad (1.103)$$

$$r(0) = 0 \ . \qquad\qquad\qquad\qquad (1.104)$$

To solve this Cauchy problem, we produce the solution to Eqs. (1.103) and (1.104) from $x = 0$ to $x = t$ . At $x = t$, we adjoin Eqs. (1.99) and (1.100), and we use the initial conditions in Eqs. (1.101) and (1.102). Continue the solution of the entire system out to $x = L$ . Then the internal intensities are known for fixed $t$ and all $x \leq L$ .

We now give a rederivation along analytical lines. Equations (1.99) - (1.104) may be derived analytically beginning with Eqs. (1.95) and (1.96) which we re-write as

$$m_t(t,x) = -m(t,x) + \frac{\lambda}{2}\Big[m(t,x) + n(t,x)\Big] , \qquad (1.105)$$

$$m(0,x) = 0 , \qquad (1.106)$$

$$n_t(t,x) = -n(t,x) + \frac{\lambda}{2}\Big[m(t,x) + n(t,x)\Big] , \qquad (1.107)$$

$$n(x,x) = 1 . \qquad (1.108)$$

Note that we shall use the subscripts t and x to denote differentiation with respect to the first and second arguments, respectively. First, differentiate throughout Eqs. (1.105) – (1.108) with respect to x . Assuming that differentiations with respect to x and t may be interchanged, the results are

$$\frac{\partial}{\partial t} m_x(t,x) = -m_x(t,x) + \frac{\lambda}{2}\Big[m_x(t,x) + n_x(t,x)\Big] ,$$
$$(1.109)$$

$$m_x(0,x) = 0 , \qquad (1.110)$$

$$-\frac{\partial}{\partial t} n_x(t,x) = -n_x(t,x) + \frac{\lambda}{2}\Big[m_x(t,x) + n_x(t,x)\Big] ,$$
$$(1.111)$$

$$\frac{d}{dx} n(x,x) = 0 . \qquad (1.112)$$

Equation (1.112) is also written

$$0 = n_t(x,x) + n_x(x,x) \tag{1.113}$$

or, using Eq. (1.107),

$$0 = n(x,x) - \frac{\lambda}{2}\Big[m(x,x) + n(x,x)\Big] + n_x(x,x) \ . \tag{1.114}$$

Introduce the function $r(x)$ ,

$$r(x) = m(x,x) \ , \tag{1.115}$$

and write Eq. (1.114) in the form

$$n_x(x,x) = -1 + \frac{\lambda}{2}[\,r(x) + 1\,] \ . \tag{1.116}$$

Compare Eqs. (1.105) - (1.108) for $m$ and $n$ against Eqs. (1.109) - (1.111) and (1.116) for $m_x$ and $n_x$ , all as functions of $t$ with $x$ held fixed. Both systems have the same linear operators and the same initial conditions. While the terminal condition of the first is unity (Eq. (1.108), that of the other system is $-1 + (\lambda/2)(r+1)$ . The solution of the latter system may thus be expressed as

$$m_x(t,x) = \Big\{-1 + \frac{\lambda}{2}\,[r(x) + 1]\Big\} \, m(t,x) \ , \tag{1.117}$$

$$n_x(t,x) = \Big\{-1 + \frac{\lambda}{2}\,[r(x) + 1]\Big\} \, n(t,x) \ . \tag{1.118}$$

These are precisely the differential Eqs. (1.109) and (1.111) which we derived earlier. Equation (1.101) follows by putting $x = t$ in Eq. (1.115), while Eq. (1.102) follows from the boundary conditions of Eq. (1.108).

In order to derive the differential equation for
$r(x)$ , differentiate on both sides of Eq. (1.115) thereby
obtaining

$$r'(x) = m_t(x,x) + m_x(x,x)$$

$$= - m(x,x) + \frac{\lambda}{2} [m(x,x) + n(x,x)]$$

$$+ \left\{ - 1 + \frac{\lambda}{2} [r(x) + 1] \right\} m(x,x)$$

$$= - r(x) + \frac{\lambda}{2} [r(x) + 1]$$

$$- r(x) + \frac{\lambda}{2} [r(x) + 1] r(x) . \qquad (1.119)$$

Finally, we have the differential equation

$$r'(x) = -2r(x) + \frac{\lambda}{2} [r(x) \; r + 1]^2 \qquad (1.120)$$

and the initial condition

$$r(0) = m(0,0) = 0 . \qquad (1.121)$$


1.7   INTERNAL INTENSITIES DUE TO INTERNAL SOURCES

When there are internal sources of particles with the
distribution $g(t)$ and no external sources, the two point
boundary value problem (for the internal intensities to the
right and left) becomes

$$\dot{v} = - v + u , \qquad v(0,L) = 0 \qquad (1.122)$$

$$- \dot{w} = - w + u , \qquad w(L,L) = 0 . \qquad (1.123)$$

where the source function  u  is

$$u = \frac{\lambda}{2} (v + w) + g .$$                                (1.124)

The Cauchy problem is given by the equations

$$v_x(t,x) + [g(x) + \frac{\lambda}{2} e(x)] \, m(t,x) ,$$              (1.125)

$$w_x(t,x) = [g(x) + \frac{\lambda}{2} e(x)] \, n(t,x) , \qquad x \geq t ,$$   (1.126)

$$e'(x) = - e(x) + \left[\frac{\lambda}{2} e(x) + g(x)\right] \left[1 + r(x)\right] ,$$   (1.127)
$$x \geq 0 ,$$

$$v(t,t) = e(t) ,$$                                                (1.128)

$$w(t,t) = 0 ,$$                                                  (1.129)

$$e(0) = 0 ,$$                                                    (1.130)

together with equations (1.99) - (1.104) for  m, n,  and  r.


## 1.8    DISCUSSION

We have considered a simple model of multiple scattering in a one-dimensional medium. We have taken physical,
analytical and computational aspects into consideration.  The
importance of some of these concepts will be made clearer in
the remainder of this book.

## EXERCISES

1.  Derive the initial value problem for internal intensities due to internal sources, Eqs. (1.122) - (1.130), by a physical approach.

2.  Write a computer program for the reflection function $r(x)$ based on Eqs. (1.13) and (1.14). Produce a family of solutions for $0 < x < 2$, for $\lambda = 0.5$. Repeat, with $\lambda = 1.0$.

3.  Write a computer program for the source function $J(t,x)$ using Eqs. (1.18) and (1.19) in addition to the equations for $r(x)$. Produce the source functions at $t = 0, 0.1, 0.2, \ldots, x$, for $x = 0.1, 0.2, \ldots, 2.0$ and $\lambda = 0.5$. Repeat, with $\lambda = 1.0$.

4.  Prove that the solution of the initial value problem for $u(t,x)$, Eqs. (1.61) - (1.66), satisfies the integral equation, Eq. (1.36).

CHAPTER 2

ISOTROPIC SCATTERING IN SLABS:
AUXILIARY PROBLEM

## 2.1    INTRODUCTION

In this chapter we shall consider various aspects of
multiple scattering in a homogeneous medium contained between
two parallel planes.  The sources are either uniform parallel
rays incident on the medium or internal isotropic sources.
Within the medium absorption and isotropic scattering occur.
We introduce the emergence probability function, the source
function, and the internal intensity function.  The first two
satisfy Fredholm integral equations, and the third satisfies
a two point boundary value problem.

But the primary objective is to obtain initial value
problems for these functions.  These Cauchy systems provide
the basis for our numerical treatment.  The reason behind this
objective is the ability of modern computers to solve large
systems of ordinary differential equations subject to a complete set of initial conditions.  Extremely high precision is
possible, provided that the differential equations are stable.
Fortunately for our imbedding approach, this is indeed the case.

31

In order to formulate initial value (Cauchy) problems
for the functions of interest, we take a rather unconventional
view of the complex multiple scattering process.  Instead of
varying the point of observation (i.e., optical depth), as is
customary, we vary the total thickness of the medium.  We imbed
the given problem in a class of problems, where the thickness
is varied and all else remains the same.  This can be done
either physically from first principles or analytically from
a classical formulation.  These alternate derivations enable
us to check the validity of our equations.

In the Cauchy problems for other than one-dimensional
(rod) problems, one encounters integro-differential equations
rather than the desired ordinary differential equations.  But
this is no obstacle.  We approximate the integrals by sums
using a Gaussian quadrature formula which yields high accuracy
at relatively low computing expense.  The new dependent
variables are the solution functions evaluated at certain values
of the angular argument.  The independent variable of the
system of approximating differential equations is the thickness of the medium.  Initial conditions are imposed at certain
values of thickness, for example, when the thickness is zero,
or when it is  t, the optical altitude of the point of observation.  This will be explained in greater detail in later sec-
tions of this chapter.

We first take up classical approaches to the multiple
scattering problem for a homogeneous finite slab.


## 2.2    EMERGENCE PROBABILITY

We shall consider radiative transfer processes in
finitely thick slabs with isotropic multiple scattering.

Consider a homogeneous absorbing and isotropically scattering medium in the form of a horizontal slab bounded by two in-finite parallel planes separated by an optical distance $x$. The properties of the medium do not depend on the altitude $t$ above the bottom. The lower boundary is assumed to be a perfect absorber so that any radiation incident on it is completely absorbed. Later, we shall analyze the cases of specular and diffuse reflecting boundaries. The upper boundary allows passage of energy in any direction.

The optical distance unit is such that in a distance $\Delta$, the fraction $\Delta + o(\Delta)$ of propagating energy is absorbed. The function $o(\Delta)$ has the property that

$$\lim_{\Delta \to 0} \frac{o(\Delta)}{\Delta} = 0 \ . \tag{2.1}$$

The fraction of absorbed energy that is re-emitted (scattered) isotropically is $\lambda$; the fraction $1 - \lambda$ is permanently absorbed and removed from the process in an elementary act of absorption.

Let $n(s)$ be the probability that no interaction (i.e., absorption) takes place when radiation passes through a distance $s$ in the slab. For this function we see that

$$n(s+\Delta) = n(s)\ (1-\Delta) + o(\Delta) \tag{2.2}$$

It follows, by letting $\Delta \to 0$, that

$$dn/ds = -\ n(s) \ . \tag{2.3}$$

Also, we have the condition that

$$n(0) = 1 . \tag{2.4}$$

From this we see that

$$n(s) = e^{-s} . \tag{2.5}$$

We shall now proceed to derive the basic integral
equation which is satisfied by the emergence probability
function  p.  It is both customary and convenient to
designate a direction by its direction cosine v  rather than
the polar angle θ.  (Note that  $|\sin \theta \ d\theta \ d\phi| = |dv \ d\phi|$.)
Let us define:

> $p(t,x,v) \ dv \ d\phi$ = the probability that a light quantum
> absorbed at altitude  t  ultimately emerges from the
> top of the slab of thickness  x,  at an angle arc
> cosine  v  within the solid angle
> $d\Omega = dv \ d\phi$ ,  for  $0 \leq t \leq x$,  $0 < x$,  $0 \leq v \leq 1$ .  (2.6)

Here,  $\phi$  is the azimuth angle.  The word "ultimately"
indicates that one or more scattering (re-emission) events
may be encountered by the light quantum.

The quantity  $p(t,x,v) \ dv \ d\phi$  may be expressed as the
sum of three terms.  The first is the probability of emergence
(directly after re-emission) with absorption but no scattering,

$$\lambda \ \frac{dv \ d\phi}{4\pi} \ e^{-(x-t)/v} . \tag{2.7}$$

The first factor,  $\lambda$,  is the probability of re-emission
into all directions.  The probability of re-emission into

the solid angle interval of interest, $dv\, d\phi$ , is then $\lambda\, dv\, d\phi/4\pi$ . The quantum of light travels to the top of the slab, with direction cosine v. The distance traversed is $(x-t)/v$, and the probability of no absorption is $n[(x-t)/v] = \exp\,[-(x-t)/v]$ .

The light quantum absorbed at t may also emerge from the top after one or more scatterings. The probability of re-emission into the direction cosine interval $(v', v' + dv')$ and in all azimuth directions is $\lambda\cdot 2\pi dv'/4\pi$. The probability of reaching altitude y, for $y > t$ , is $\exp[-(y-t)/v']$. The probability of an interaction between y and $y + dy$ is $dy/v'$. With the quantum being absorbed at y, the probability of ultimate emergence in the direction interval $dvd\phi$ is $p(y,x,v)\, dvd\phi$. These processes contribute the term

$$\int_0^1 \int_t^x \lambda\, \frac{2\pi dv'}{4\pi}\, e^{-(y-t)/v'}\, \frac{dy}{v'}\, p(y,x,v)\, dvd\phi\ ,$$

where we integrate over all $v'$ from 0 to 1. Similarly, the quantum absorbed at t may be re-emitted into a down-going direction whose direction cosine is $v'$ relative to the down-going vertical. It may travel to altitude y , for $y < t$, be absorbed there, and ultimately emerge within the desired solid angle.

The mathematical expression for this probability is

$$\int_0^1 \int_0^t \lambda\, \frac{2\pi dv'}{4\pi}\, e^{-(t-y)/v'}\, \frac{dy}{v'}\, p(y,x,v)\, dvd\phi\ .$$

Combining these three terms, we obtain the equation

$$p(t,x,v)\ dvd\phi\ =\ \frac{\lambda}{4\pi}\ dvd\phi\ e^{-(x-t)/v}$$

$$+\ \int_0^1 \int_0^x \frac{2\pi dv'}{4\pi}\ e^{-|t-y|/v'}\ \frac{dy}{v'}\ p(y,x,v)\ dvd\phi\ ,$$

$$\text{(2.8)}$$

which simplifies to

$$p(t,x,v)\ =\ \frac{\lambda}{4\pi}e^{-(x-t)/v}$$

$$+\ \frac{\lambda}{2}\ \int_0^x \int_0^1\ e^{-|t-y|/v'}\ \frac{dv'}{v'}\ p(y,x,v)\ dy\ .$$

$$\text{(2.9)}$$

The first exponential integral function is defined by the integral

$$E_1(r)\ =\ \int_0^1\ e^{-r/v'}\ \frac{dv'}{v'}\ ,\qquad\qquad r>0\ .\ \text{(2.10)}$$

The auxiliary Fredholm integral equation thus takes the form

$$p(t,x,v)\ =\ \frac{\lambda}{4\pi}\ e^{-(x-t)/v}$$

$$+\ \frac{\lambda}{2} \int_0^x E_1(|t-y|)\ p(y,x,v)\ dy\ ,$$

$$0\leq t\leq x,\quad 0\leq v\leq 1\ .\ \text{(2.11)}$$

Consider now the sources of energy in the slab. Let the rate of emission per unit solid angle in a certain volume be $g(y)dy$, where the emitting volume extends from altitude $y$ to $y + dy$. The rate of absorption is expressed as $4\pi g(y)/\lambda$. Then the rate at which energy emerges from the top of the slab going into the solid angle interval $(v,v+dv)$

and $(\phi, \phi + d\phi)$, and passing through a unit of area taken normal to the propagation direction, is

$$e(x,v) \; dv \; d\phi \; = \frac{1}{v} \int_0^x \; \frac{4\pi}{\lambda} \; g(y) \; dy \; p(y,x,v) \; dv \; d\phi \; .$$
$$(2.12)$$

The emergence probability function thus yields the emergent intensity due to an arbitrary distribution of internal sources,

$$e(x,v) \; = \frac{1}{v} \int_0^x \; \frac{4\pi}{\lambda} \; g(y) \; p(y,x,v) \; dy \; ,$$
$$0 \leq v \leq 1, \quad 0 \leq x \; . \quad (2.13)$$

## 2.3    MONODIRECTIONAL ILLUMINATION - INTEGRAL EQUATION

When a plane-parallel atmosphere is illuminated by uniform parallel rays of illumination, the source function satisfies an integral equation like the one for the probability of emergence. The two functions differ only by a constant factor, as will be seen. This source function is another basic function because in terms of it, the source function for the case of an arbitrary linear combination of exponentially distributed sources can be expressed as an integral. Furthermore, internal intensities and reflection and transmission functions are certain integrals of source functions.

Let us consider the absorption and scattering processes involved when energy passes through the medium. Intensity at a point in the medium, in a given direction, is the rate of energy flow per unit of area oriented normal to the given direction per unit of solid angle. Let $I(s,v)$ denote the intensity at a point designated by the coordinate $s$ and

propagating with direction arccos v measured from the upward-directed normal. Consider a small cylindrical element
of the medium, with base area $d\sigma$ and height $ds$, the
faces being perpendicular to the flow direction. The energy
flowing through the face located at s, per unit solid angle
and per unit time is, $I(s,v)d\sigma$; that flowing through the
face at s+ds is $I(s+ds,v)d\sigma$. The latter quantity is
expressed as

$$I(s+ds,v) \ d\sigma = I(s,v)d\sigma - ds \cdot I(s,v)d\sigma$$

$$+ \ J(s)dsd\sigma + o(ds) \ . \qquad (2.14)$$

The second term on the right is due to absorption, and the
third term is due to scattering (and emission, if any). The
source function $J(s)$ is the rate of production of scattered
radiation per unit volume per unit solid angle at s.
Divide through Eq. (2.14) by $dsd\sigma$. We can then write

$$\frac{I(s+ds,v) - I(s,v)}{ds} = - \ I(s,v) + J(s) + \frac{o(ds)}{dsd\sigma} \cdot (2.15)$$

Let ds tend to 0 to obtain the differential equation

$$\frac{dI(s,v)}{ds} = - \ I(s,v) + J(s) \ . \qquad (2.16)$$

The transfer equation, Eq. (2.16), tells what happens as we
follow the radiation along a beam. The mathematical problem
is still incomplete at this stage, for we have not specified
the primary source of energy, nor the boundary conditions.

   Let us now describe the source of radiation. Uniform
parallel rays of net flux $\pi$ are incident on the top of a

plane-parallel atmosphere.  This means that the incident
energy passing through a unit square oriented perpendicularly
to the incident radiation per unit of time is  $\pi$.  The acute
angle between the inward drawn normal and the direction of
the incident radiation is arccos  u .

Let  $I(t,v,x,u)$  denote the intensity of the diffuse
radiation at the altitude  t  in a direction making an angle
arccos    v  with respect to the upward normal  $(-1 \leq v \leq 1)$ ,
the incident radiation being as described above.  Diffuse
radiation refers to the radiation which has interacted at
least once with the medium and survived.  Let the source
function  $J(t,x,v)$  be the rate of production of scattered
radiation per unit volume per unit solid angle at the altitude
t .  We shall now derive equations for these functions.

In view of Eq. (2.16), we may write

$$v \, \frac{dI(t,v,x,u)}{dt} = - \, I(t,v,x,u) + J(t,x,u) \; . \qquad (2.17)$$

The boundary conditions are that no diffuse radiation is
incident or reflected inwards at the boundaries; therefore,

$$I(x,v,x,u) = 0 \; , \qquad\qquad\qquad -1 \leq v < 0 \; , \quad (2.18)$$

$$I(0,v,x,u) = 0 \; , \qquad\qquad\qquad 0 < v \leq 1 \; . \quad (2.19)$$

Regard the source function  J  in Eq. (2.17) as a forcing
function.  The formal solution for  I  in terms of  J  is

$$I(t,v,x,u) = \begin{cases} \dfrac{1}{v}\displaystyle\int_0^t J(y,x,u)\, e^{-(t-y)/v}\, dy\,, & 0<v\le 1\,, \\[2em] -\dfrac{1}{v}\displaystyle\int_t^x J(y,x,u)\, e^{-(t-y)/v}\, dy\,, & -1\le v<0. \end{cases}$$
$$(2.20)$$

The next task is to find an equation for  J.  First, though, we know that  J  may be expressed in terms of  I  as follows.  The rate of scattering of energy per unit solid angle in a cylinder of base area  $d\sigma$  and height  dt  located between  (t,t+dt)  is

$$J(t,x,u)\, dt\, d\sigma = \int_{-1}^{+1} vI(t,v,x,u)d\sigma \cdot \frac{dt}{v} \cdot \frac{\lambda}{4\pi} \cdot 2\pi dv$$
$$+ \pi u \cdot e^{-(x-t)/u} \cdot \frac{dt}{u} \cdot \frac{\lambda}{4\pi} \cdot d\sigma + \dots$$
$$(2.21)$$

On the right hand side is an integral on  v  from  -1  to  +1.  The factor  $vId\sigma$  is the energy passing through a horizontal face with area  $d\sigma$ .  The fraction of this energy that interacts is  dt/v,  the fraction which is scattered into a unit of solid angle is  $\lambda/4\pi$.  The energy was original- ly flowing with all azimuths  $(2\pi)$  and with direction cosine in the interval  (v,v+dv).  The exponential term is due to the reduced incident radiation.  The energy per unit horizontal area incident at the top per unit time is  $\pi u$.  This energy propagates to altitude  t+dt  arriving with strength  $\pi u\exp[-(x-t)/u]$.  The fraction  dt/u  is absorbed and  $\lambda/4\pi$  is the fraction re-emitted.  The factor  $d\sigma$  is present be- cause we are considering the area  $d\sigma$  rather than unit area. The equation for  J  in terms of  I  becomes

$$J(t,x,u) = \frac{\lambda}{2} \int_{-1}^{+1} I(t,v,x,u) \, dv$$

$$+ \frac{\lambda}{4} e^{-(x-t)/u} .$$
(2.22)

Eliminating I from Eqs. (2.20) and (2.22), we obtain

$$J(t,x,u) = \frac{\lambda}{2} \int_0^{+1} \int_0^t J(y,x,u) e^{-(t-y)/v} \, dy \, dv$$

$$+ \frac{\lambda}{2} \int_{-1}^0 \int_t^x J(y,x,u) e^{-(t-y)/v} \, dy \, dv$$

$$+ \frac{\lambda}{4} e^{-(x-t)/u} ,$$
(2.23)

or

$$J(t,x,u) = \frac{\lambda}{4} e^{-(x-t)/u}$$

$$+ \frac{\lambda}{2} \int_0^x \int_0^1 e^{-|t-y|/v} \frac{dv}{v} J(y,x,u) \, dy ,$$
(2.24)

or, finally,

$$J(t,x,u) = \frac{\lambda}{4} e^{-(x-t)/u}$$

$$+ \frac{\lambda}{2} \int_0^x E_1(|t-y|) J(y,x,u) \, dy , \quad 0 \le t \le x.$$
(2.25)

Equation (2.25) is the integral equation for the source function J .

Comparison of the integral equations for p and J reveals that

$$p(t,x,v) = \frac{1}{\pi} J(t,x,v) \ . \qquad\qquad (2.26)$$

We have thus seen that two classical descriptions of the basic problem of radiative transfer are the integral equations and the two-point boundary-value problem. These are generally unsuitable for numerical computation. But there is a third description as an initial value problem. This modern approach is indeed computationally successful.

## 2.4     MONODIRECTIONAL ILLUMINATION - CAUCHY SYSTEM FOR SOURCE FUNCTION

We will derive a complete initial value problem for the source function. The basic concept is to regard the slab thickness as a variable — the independent variable. We derive equations which show how the source function at a fixed internal point varies as the thickness increases. Auxiliary functions are introduced to complete the system. These, too, are functions of thickness. We obtain a system of integrodifferential equations and a complete set of initial conditions. We then show how to solve the approximate system of ordinary differential equations with initial conditions. Finally, numerical results are presented.

Consider a slab of arbitrary thickness $x$, and another slab of slightly differing thickness $x + \Delta$. The source function for the slab of thickness $x + \Delta$ may be expressed

$$J(t,x+\Delta,u) = \left[1 - \frac{\Delta}{u}\right] J(t,x,u)$$

$$+ \; J(x+\Delta,x+\Delta,u)\Delta \int_0^1 \frac{1}{\pi} J(t,x,u')2\pi du'+o(\Delta).$$

$$(2.27)$$

The first factor of the first term in the right hand side, $1-\Delta/u$, accounts for the diminution in the incident radiation as it passes through the first layer of thickness $\Delta$ at the top of the slab of thickness $x$, and the second factor is the rate of scattering due to the reduced source incident on the remainder of the slab which is then of thickness $x$. The second term accounts for the contribution due to scattering at the top of the slab of thickness $x + \Delta$, in a cylinder of unit base area and height $\Delta$ with rate $J(x+\Delta,x+\Delta,u)$. Of this amount the fraction $2\pi du'$ is scattered into the direction cosine interval $(u',u'+du')$. This scattering produces omnidirectional sources at the top. The effect of these sources can be had by summing up the separate effects due to components with direction cosine of incidence $u'$. The division by $u'$ converts the source strength into units of net flux $\pi$ per unit normal area. Then the scattering at $t$ due to such a source is $J(t,x,u')$. Other processes involve $\Delta^2,\Delta^3,\dots$ and are included in the term $o(\Delta)$.

The quantity $J(x+\Delta, x+\Delta, u)$ may be expressed as a power series in $\Delta$,

$$J(x+\Delta, x+\Delta, u) = J(x,x,u) + \frac{d}{dx} J(x,x,u) \; \Delta + o(\Delta) \; .$$
$$(2.28)$$

The second term on the right hand side of Eq. (2.27) may be then written

$$J(x,x,u)\Delta \int_0^1 \frac{1}{\pi u'} J(t,x,u') \; 2\pi du' + o(\Delta) \; . \qquad (2.29)$$

Equation (2.27) becomes

$$J(t,x+\Delta, u) = \left[ 1 - \frac{\Delta}{u} \right] J(t,x,u)$$

$$+ 2J(x,x,u) \int_0^1 J(t,x,u') \frac{du'}{u'} \Delta + o(\Delta) \; ,$$
$$(2.30)$$

and can be rewritten

$$\frac{J(t,x+\Delta,u) - J(t,x,u)}{\Delta} = \frac{-1}{u} J(t,x,u)$$

$$+ 2J(x,x,u) \int_0^1 J(t,x,u') \frac{du'}{u'}$$

$$+ \frac{o(\Delta)}{\Delta} \; . \qquad (2.31)$$

In the limit as $\Delta$ tends to zero this becomes

$$J_x(t,x,u) = - \frac{1}{u} J(t,x,u) + 2J(x,x,u) \int_0^1 J(t,x,u')du'/u' \; ,$$
$$x \geq t \; , \qquad (2.32)$$

which is the desired differential equation. This equation
is obtainable from the integral equation (2.25) by differentiation with respect to x, the thickness.

The differential Eq. (2.32) is derived analytically
by differentiation of both sides of the integral Eq. (2.25)
which yields

$$J_x(t,x,u) = -\frac{1}{u} \frac{\lambda(x)}{4} e^{-(x-t)/u} + \frac{\lambda(x)}{2} E_1(x-t) \ J(x,x,u)$$

$$+ \frac{\lambda(x)}{2} \int_0^x E_1(|t-y|) \ J_x(y,x,u) \ dy \ . \quad (2.33)$$

This integral equation for the function $J_x(t,x,u)$ can be
solved in the form of Eq. (2.32) by application of the superposition property of linear integral equations.

Note that the right hand side of Eq. (2.32) depends
on $J(x,x,u)$. We shall next show how to determine this
function.

Consider the source functions at the top and bottom,

$$A(x,u) = J(x,x,u) \ , \quad\quad\quad\quad\quad\quad\quad\quad (2.34)$$

$$B(x,u) = J(0,x,u). \quad\quad\quad\quad\quad\quad\quad\quad (2.35)$$

By putting t = 0 in the differential equation for $J(t,x,u)$
one obtains

$$B_x(x,u) = -\frac{1}{u} B(x,u) + A(x,u) \ 2\int_0^1 B(x,u') \ du'/u'$$
$$x \geq 0 \ . \quad (2.36)$$

A differential equation for $A(x,z)$ is obtained by adding
a thickness $\Delta$ to the bottom of the slab of thickness x.

We write

$$A(x+\Delta,u) = A(x,u) + B(x,u)\Delta \ \cdot$$

$$\cdot \int_0^1 \frac{1}{\pi u'} B(x,u') \ 2\pi du' + o(\Delta) \ . \qquad (2.37)$$

We let $\Delta$ tend to 0 and obtain the equation

$$A_x(x,u) = B(x,u) \ 2\int_0^1 B(x,u') \ du'/u' \qquad x\geq 0 \ \cdot (2.38)$$

To derive the initial condition on $B(x,z)$ consider the rate of production of scattered energy in a volume with unit base area and height $\Delta$, per unit solid angle, in a slab of thickness $\Delta$. It is

$$B(\Delta,u)\Delta \quad = \quad \pi u \ \frac{\Delta}{u} \cdot \frac{\lambda}{4\pi} + o(\Delta) \ . \qquad (2.39)$$

Divide through by $\Delta$ and let $\Delta$ tend to 0. This yields

$$B(0,u) = \frac{\lambda}{4} \ . \qquad (2.40)$$

Similarly, the initial condition on A is

$$A(0,u) = \frac{\lambda}{4} \ . \qquad (2.41)$$

The X and Y functions of radiative transfer have initial conditions unity and may be introduced as follows:

$$J(x,x,u) = A(x,u) = \frac{\lambda}{4} X(x,u) \ , \qquad (2.42)$$

$$J(0,x,u) = B(x,u) = \frac{\lambda}{4} Y(x,u) . \qquad (2.43)$$

A complete Cauchy system for the family of source functions $J(t,x,u)$ consists of the equations

$$J_x(t,x,u) = -\frac{1}{u} J(t,x,u) + \frac{\lambda}{2} X(x,u) \int_0^1 J(t,x,u') \frac{du'}{u'} ,$$

$$x \geq t , \qquad (2.44)$$

$$X_x(x,u) = \frac{\lambda}{2} Y(x,u) \int_0^1 Y(x,u') \frac{du'}{u'} , \qquad x \geq 0 , \qquad (2.45)$$

$$Y_x(x,u) = -\frac{1}{u} Y(x,u) + \frac{\lambda}{2} X(x,u) \int_0^1 Y(x,u') \frac{du'}{u'} ,$$

$$x \geq 0 . \qquad (2.46)$$

$$J(t,t,u) = \frac{\lambda}{4} X(t,u) , \qquad (2.47)$$

$$X(0,u) = 1 , \qquad (2.48)$$

$$Y(0,u) = 1 , \qquad (2.49)$$

for $0 < u \leq 1$ .

Suppose that our aim is to calculate the family of source functions, $J(t,x,u)$, for the fixed points $t = t_1$, $t_2, \ldots, t_M, x$, $(0 < t_1 < t_2 \ldots < t_M < x)$, for $0 < x \leq x_{max}$ , and for $0 < u \leq 1$. Let us describe how this is to be done through use of Eqs. (2.44) - (2.49). We first note that if the sole task were to produce the functions $X$ and $Y$ , we simply have to solve the differential Eqs. (2.45) and (2.46) simultaneously for $x > 0$, subject to the initial conditions of Eqs. (2.48) and (2.49). The physical picture is that of starting with a slab of zero thickness, and allowing the slab to grow thicker until $x = x_{max}$. Knowing the functions X and Y, we of course know the source functions at the top and bottom of the slab, through Eqs. (2.42) and (2.43).

Let us next indicate how to determine the source function at the interior point whose altitude is $t_1$, for all $x < x_{max}$ and all u. Begin again with the slab of zero thickness and X and Y set equal to one (Eqs. (2.48) and (2.49)). Solve Eqs. (2.45) and (2.46) to $x = t_1$. We introduce the function $J(t_1,x,u)$ as a function of x, with $t_1$ being fixed and u taking on all values $0 < u \le 1$. Its initial condition is given by Eq. (2.47) with $t = t_1$, in terms of the known X function. Its differential Eq. (2.44), with $t = t_1$, is adjoined to the system of equations for X and Y, and the entire system is simultaneously solved for $t_1 < x < x_{max}$.

To produce the source function at the M interior points $t = t_1, t_2, \ldots, t_M$, we need merely to use the appropriate initial condition, Eq. (2.47), with $t = t_i$, whenever x reaches the value $t_i$, for $i = 1, 2, \ldots, M$; and in addition we adjoin the appropriate differential Eq. (2.44) with $t = t_i$. The system is enlarged whenever $x = t_i$, and the complete system is solved out to $x = x_{max}$. Picture a slab of thickness zero and some equations for X and Y. Let x increase until it is $t_1$, when equations for $J(t_1,x,u)$ are introduced. The enlarged system is solved as the thickness x gets greater. At $x = t_2$, we repeat the process for $J(t_2,x,u)$, and so on. Finally, the thickness has reached $x = t_{max}$ and the problem has, in principle, been solved. In the next section, we explain how this is done computationally.

In this way, we sweep out a whole family of source functions and we obtain a valuable parameter study in thickness, for various internal points and for all incident angles.


## 2.5 COMPUTATIONAL METHOD AND RESULTS

In order to deal with the above initial value
problem for integro-differential equations, we make the
approximation that integrals can be replaced by sums according to the quadrature formula,

$$\int_0^1 f(z)\, dz \cong \sum_{i=1}^{N} f(z_i) w_i \ ,$$

where $z_1, z_2, \ldots, z_N$ are the points at which the integrand
is evaluated, and $w_1, w_2, \ldots, w_N$ are the corresponding
Christoffel weights. We use a Gaussian quadrature formula
of order N. This formula is exact for polynomials up to
degree 2N-1. For example, for N=7, the quadrature is
exact if the polynomial is of degree 13; this is sufficiently
accurate for many practical purposes. The points
$z_1, z_2, \ldots, z_N$ are the roots of the shifted Legendre polynomial, $P_N(1-2x)$, which are in the interval $(0,1)$. These
roots and weights have been tabulated for various N.
Table 2.1 and Appendix A give them for N=7. For other
values of N, see Ref. 2. Table 2.1 also gives the angles,
arccos $z_i$.

Table 2.1

Points, Weights and Angles for Gaussian
Quadrature with  N = 7

$$\int_0^1 f(x) \, dx \cong \sum_{i=1}^{N} f(z_i) w_i$$

| i | Points $z_i$ | Weights $w_i$ | Angles (deg) arccos $z_i$ |
|---|---|---|---|
| 1 | .025446046 | .064742484 | 88.5419 |
| 2 | .12923441 | .13985269 | 82.5746 |
| 3 | .29707742 | .19091502 | 72.7178 |
| 4 | .50000000 | .20897958 | 60.0000 |
| 5 | .70292258 | .19091502 | 45.3380 |
| 6 | .89076559 | .13985269 | 29.4523 |
| 7 | .97455396 | .064742484 | 12.9531 |

In the Cauchy problem, the integrals are replaced by sums according to a Gaussian quadrature formula of order  N. The dependent variables are

$$J_i(t,x) = J(t,x,z_i) \ , \tag{2.50}$$

$$X_i(x) = X(x,z_i) \ , \tag{2.51}$$

$$Y_i(x) = Y_i(x,z_i) \ , \qquad i=1,2,\ldots,N. \tag{2.52}$$

The system of ordinary differential equations is

$$J_i' = -\frac{1}{z_i} J_i + \frac{\lambda}{2} X_i \sum_{j=1}^{N} J_j \frac{w_j}{z_j} , \qquad x \geq t , \qquad (2.53)$$

$$X_i' = \frac{\lambda}{2} Y_i \sum_{j=1}^{N} Y_j \frac{w_j}{z_j} , \qquad x \geq 0 , \qquad (2.54)$$

$$Y_i' = -\frac{1}{z_i} Y_i + \frac{\lambda}{2} X_i \sum_{j=1}^{N} Y_j \frac{w_j}{z_j} , \qquad x \geq 0 , \qquad (2.55)$$

subject to the initial conditions

$$J_i(t,t) = \frac{\lambda}{4} X_i(t) , \qquad\qquad\qquad (2.56)$$

$$X_i(0) = 1 , \qquad\qquad\qquad (2.57)$$

$$Y_i(0) = 1 , \qquad\qquad\qquad (2.58)$$

for   i = 1,2,...,N .                                              (2.59)

A table of  X  and  Y  functions is given in Table 2.2 for $\lambda$ = 1.0, and thicknesses  x = 0,0.1,0.2,...,2.5.  The seven columns refer to the seven incident angles, arccos $z_i$, for  i = 1,2,...,7  as given in Table 2.1.  Two rows of data are given for each thickness:  the first gives values of  $X_i$, the second,  $Y_i$,  i = 1,2,...,7.  For example we see from the table that  $Y_1(0.1)$ = 5.22771E-02 = 5.22771 x 10$^{-2}$.  This is a portion of actual computer output for calculation of X and Y  functions via the initial value problem.

Source functions are presented in Tables 2.3 - 2.6 for the albedoes and thicknesses  $\lambda$ = 0.3, x = 5.0; $\lambda$ = 0.5,

## Table 2.2

X AND Y FUNCTIONS,   ALBEDO = 1.0000

| THICKNESS | 1 | 2 | 3 | 4 | 5 | 6 |
|---|---|---|---|---|---|---|
| 0. | 1.00000E 00 | 1.00000E 00 | 1.00000E 00 | 1.00000E 00 | 1.00000E 00 | 1.00000E 00 | 1.00000E 00 |
|  | 1.00000E 00 | 1.00000E 00 | 1.00000E 00 | 1.00000E 00 | 1.00000E 00 | 1.00000E 00 | 1.00000E 00 |
| 0.1000 | 1.05294E 00 | 1.12251E 00 | 1.14461E 00 | 1.15277E 00 | 1.15643E 00 | 1.15822E 00 | 1.15902E |
|  | 5.22771E-02 | 5.70326E-01 | 8.51605E-01 | 9.66939E-01 | 1.02049E 00 | 1.04699E 00 | 1.05904 |
| 0.2000 | 1.05625E 00 | 1.16641E 00 | 1.22118E 00 | 1.24456E 00 | 1.25563E 00 | 1.26116E 00 | 1.26369E |
|  | 2.54532E-02 | 3.40591E-01 | 7.06569E-01 | 8.98214E-01 | 9.95467E-01 | 1.04555E 00 | 1.06876E |
| 0.3000 | 1.05814E 00 | 1.18911E 00 | 1.27424E 00 | 1.31538E 00 | 1.33586E 00 | 1.34633E 00 | 1.35116E 00 |
|  | 2.12553E-02 | 2.23725E-01 | 5.91162E-01 | 8.30273E-01 | 9.62275E-01 | 1.03290E 00 | 1.06621E 00 |
| 0.4000 | 1.05957E 00 | 1.20274E 00 | 1.31323E 00 | 1.37252E 00 | 1.40344E 00 | 1.41960E 00 | 1.42714E 00 |
|  | 1.87356E-02 | 1.62374E-01 | 5.00625E-01 | 7.66827E-01 | 9.25945E-01 | 1.01426E 00 | 1.05663E 00 |
| 0.5000 | 1.06071E 00 | 1.21200E 00 | 1.34306E 00 | 1.41992E 00 | 1.46174E 00 | 1.48406E 00 | 1.49459E 00 |
|  | 1.69278E-02 | 1.28590E-01 | 4.29842E-01 | 7.09039E-01 | 8.88883E-01 | 9.92276E-01 | 1.04273E 00 |
| 0.6000 | 1.06165E 00 | 1.21893E 00 | 1.36663E 00 | 1.46003E 00 | 1.51287E 00 | 1.54165E 00 | 1.55535E 00 |
|  | 1.55453E-02 | 1.08768E-01 | 3.74430E-01 | 6.57095E-01 | 8.52384E-01 | 9.68519E-01 | 1.02612E 00 |
| 0.7000 | 1.06248E 00 | 1.22448E 00 | 1.38576E 00 | 1.49448E 00 | 1.55826E 00 | 1.59364E 00 | 1.61064E 00 |
|  | 1.44380E-02 | 9.62122E-02 | 3.30839E-01 | 6.10718E-01 | 8.17129E-01 | 9.43924E-01 | 1.00782E 00 |
| 0.8000 | 1.06319E 00 | 1.22914E 00 | 1.40166E 00 | 1.52443E 00 | 1.59889E 00 | 1.64094E 00 | 1.66133E 00 |
|  | 1.35202E-02 | 8.75825E-02 | 2.96285E-01 | 5.69436E-01 | 7.83457E-01 | 9.19073E-01 | 9.88463E-01 |
| 0.9000 | 1.06381E 00 | 1.23316E 00 | 1.41513E 00 | 1.55073E 00 | 1.63554E 00 | 1.68424E 00 | 1.70806E 00 |
|  | 1.27398E-02 | 8.11830E-02 | 2.68628E-01 | 5.32715E-01 | 7.51518E-01 | 8.94338E-01 | 9.68501E-01 |
| 1.0000 | 1.06437E 00 | 1.23671E 00 | 1.42674E 00 | 1.57403E 00 | 1.66878E 00 | 1.72406E 00 | 1.75132E 00 |
|  | 1.20637E-02 | 7.61310E-02 | 2.46243E-01 | 5.00032E-01 | 7.21352E-01 | 8.69961E-01 | 9.48239E-01 |
| 1.1000 | 1.06488E 00 | 1.23988E 00 | 1.43689E 00 | 1.59483E 00 | 1.69908E 00 | 1.76083E 00 | 1.79153E 00 |
|  | 1.14696E-02 | 7.19512E-02 | 2.27902E-01 | 4.70898E-01 | 6.92939E-01 | 8.46104E-01 | 9.27898E-01 |
| 1.2000 | 1.06536E 00 | 1.24274E 00 | 1.44588E 00 | 1.61352E 00 | 1.72683E 00 | 1.79491E 00 | 1.82900E 00 |
|  | 1.09414E-02 | 6.83759E-02 | 2.12680E-01 | 4.44872E-01 | 6.66223E-01 | 8.22870E-01 | 9.07642E-01 |
| 1.3000 | 1.06575E 00 | 1.24535E 00 | 1.45392E 00 | 1.63042E 00 | 1.75233E 00 | 1.82658E 00 | 1.86403E 00 |
|  | 1.04675E-02 | 6.52461E-02 | 1.99880E-01 | 4.21562E-01 | 6.41129E-01 | 8.00329E-01 | 8.87592E-01 |
| 1.4000 | 1.06614E 00 | 1.24774E 00 | 1.46118E 00 | 1.64578E 00 | 1.77586E 00 | 1.85609E 00 | 1.89684E 00 |
|  | 1.00391E-02 | 6.24613E-02 | 1.88975E-01 | 4.00626E-01 | 6.17570E-01 | 7.78522E-01 | 8.67837E-01 |
| 1.5000 | 1.06649E 00 | 1.24993E 00 | 1.46780E 00 | 1.65983E 00 | 1.79764E 00 | 1.88367E 00 | 1.92766E 00 |
|  | 9.64924E-03 | 5.99542E-02 | 1.79566E-01 | 3.81764E-01 | 5.95456E-01 | 7.57470E-01 | 8.48445E-01 |
| 1.6000 | 1.06682E 00 | 1.25197E 00 | 1.47386E 00 | 1.67272E 00 | 1.81785E 00 | 1.90949E 00 | 1.95664E 00 |
|  | 9.29234E-03 | 5.76767E-02 | 1.71351E-01 | 3.64715E-01 | 5.74695E-01 | 7.37181E-01 | 8.29465E-01 |
| 1.7000 | 1.06712E 00 | 1.25385E 00 | 1.47944E 00 | 1.68461E 00 | 1.83666E 00 | 1.93371E 00 | 1.98396E 00 |
|  | 8.96400E-03 | 5.55933E-02 | 1.64098E-01 | 3.49252E-01 | 5.55198E-01 | 7.17651E-01 | 8.10930E-01 |
| 1.8000 | 1.06740E 00 | 1.25561E 00 | 1.48461E 00 | 1.69561E 00 | 1.85422E 00 | 1.95649E 00 | 2.00975E 00 |
|  | 8.66059E-03 | 5.36763E-02 | 1.57628E-01 | 3.35181E-01 | 5.36876E-01 | 6.98871E-01 | 7.92866E-01 |
| 1.9000 | 1.06767E 00 | 1.25725E 00 | 1.48942E 00 | 1.70583E 00 | 1.87065E 00 | 1.97794E 00 | 2.03413E 00 |
|  | 8.37909E-03 | 5.19039E-02 | 1.51803E-01 | 3.22331E-01 | 5.19648E-01 | 6.80824E-01 | 7.75286E-01 |
| 2.0000 | 1.06792E 00 | 1.25879E 00 | 1.49391E 00 | 1.71535E 00 | 1.88604E 00 | 1.99817E 00 | 2.05721E 00 |
|  | 8.11700E-03 | 5.02582E-02 | 1.46515E-01 | 3.10557E-01 | 5.03434E-01 | 6.63491E-01 | 7.58200E-01 |
| 2.1000 | 1.06815E 00 | 1.26023E 00 | 1.49812E 00 | 1.72426E 00 | 1.90051E 00 | 2.01729E 00 | 2.07909E 00 |
|  | 7.87219E-03 | 4.87246E-02 | 1.41679E-01 | 2.99732E-01 | 4.88162E-01 | 6.46848E-01 | 7.41610E-01 |
| 2.2000 | 1.06837E 00 | 1.26159E 00 | 1.50207E 00 | 1.73260E 00 | 1.91413E 00 | 2.03538E 00 | 2.09987E 00 |
|  | 7.64285E-03 | 4.72907E-02 | 1.37228E-01 | 2.89745E-01 | 4.73763E-01 | 6.30872E-01 | 7.25515E-01 |
| 2.3000 | 1.06858E 00 | 1.26287E 00 | 1.50578E 00 | 1.74044E 00 | 1.92698E 00 | 2.05253E 00 | 2.11961E 00 |
|  | 7.42744E-03 | 4.59460E-02 | 1.33109E-01 | 2.80502E-01 | 4.60173E-01 | 6.15536E-01 | 7.09912E-01 |
| 2.4000 | 1.06877E 00 | 1.26408E 00 | 1.50929E 00 | 1.74783E 00 | 1.93911E 00 | 2.06879E 00 | 2.13840E 00 |
|  | 7.22464E-03 | 4.46818E-02 | 1.29277E-01 | 2.71921E-01 | 4.47333E-01 | 6.00815E-01 | 6.94792E-01 |
| 2.5000 | 1.06896E 00 | 1.26523E 00 | 1.51261E 00 | 1.75480E 00 | 1.95060E 00 | 2.08425E 00 | 2.15629E 00 |
|  | 7.03328E-03 | 4.34903E-02 | 1.25698E-01 | 2.63930E-01 | 4.35188E-01 | 5.86683E-01 | 6.80147E-01 |

$x = 10$; $\lambda = 0.9$, $x = 20$; $\lambda = 1.0$, $x = 30$. Each line of data represents the depth as indicated (depth = $x-t$), and the values are the source functions for the seven angles of incidence, arccos $z_i$, $i = 1,2,\ldots,7$.

Let us explain how these tables were produced. The numerical integration is carried out through a Fortran program for a fourth-order Adams-Moulton method. The value of the step size used is $h = 0.005$.

Table 2.3

Source Functions for Albedo 0.3, Thickness 5

ALBEDO =0.3000     THICKNESS = 5.0000

| DEPTH | J(1) | J(2) | J(3) | J(4) | J(5) | J(6) | J(7) |
|---|---|---|---|---|---|---|---|
| 0. | 7.61188E-02 | 7.85250E-02 | 8.06983E-02 | 8.23169E-02 | 8.34111E-02 | 8.40876E-02 | 8.44348E-02 |
| 0.5000 | 2.29722E-04 | 3.15939E-03 | 1.87018E-02 | 3.56806E-02 | 4.73932E-02 | 5.43857E-02 | 5.78693E-02 |
| 1.0000 | 1.00581E-04 | 6.69796E-04 | 4.80078E-03 | 1.49725E-02 | 2.54289E-02 | 3.29626E-02 | 3.70651E-02 |
| 1.5000 | 5.00890E-05 | 3.08519E-04 | 1.52827E-03 | 6.38839E-03 | 1.35119E-02 | 1.96715E-02 | 2.33380E-02 |
| 2.0000 | 2.63642E-05 | 1.59920E-04 | 6.13041E-04 | 2.80412E-03 | 7.16287E-03 | 1.16484E-02 | 1.45620E-02 |
| 2.5000 | 1.43221E-05 | 8.62229E-05 | 2.90849E-04 | 1.27523E-03 | 3.80027E-03 | 6.86498E-03 | 9.03321E-03 |
| 3.0000 | 7.93769E-06 | 4.75588E-05 | 1.51445E-04 | 6.03059E-04 | 2.02130E-03 | 4.03264E-03 | 5.57929E-03 |
| 3.5000 | 4.45810E-06 | 2.66215E-05 | 8.25662E-05 | 2.96532E-04 | 1.07853E-03 | 2.36192E-03 | 3.43222E-03 |
| 4.0000 | 2.52264E-06 | 1.50266E-05 | 4.59877E-05 | 1.50915E-04 | 5.76697E-04 | 1.37722E-03 | 2.09949E-03 |
| 4.5000 | 1.42243E-06 | 8.45682E-06 | 2.56823E-05 | 7.84584E-05 | 3.06665E-04 | 7.92948E-04 | 1.26632E-03 |
| 5.0000 | 7.46246E-07 | 4.42999E-06 | 1.33845E-05 | 3.87322E-05 | 1.51708E-04 | 4.21824E-04 | 7.04879E-04 |

Table 2.4

Source Functions for Albedo 0.5, Thickness 10

ALBEDO =0.5000    THICKNESS = 10.0000

| DEPTH | J(1) | J(2) | J(3) | J(4) | J(5) | J(6) | J(7) |
|---|---|---|---|---|---|---|---|
| 0. | 1.28289E-01 | 1.35744E-01 | 1.42888E-01 | 1.48467E-01 | 1.52368E-01 | 1.54834E-01 | 1.56116E-01 |
| 0.5000 | 8.14272E-04 | 8.15287E-03 | 3.94372E-02 | 7.35127E-02 | 9.75279E-02 | 1.12145E-01 | 1.19519E-01 |
| 1.0000 | 3.97152E-04 | 2.56262E-03 | 1.26258E-02 | 3.45147E-02 | 5.67113E-02 | 7.27658E-02 | 8.15519E-02 |
| 1.5000 | 2.15513E-04 | 1.33053E-03 | 5.13706E-03 | 1.66124E-02 | 3.24595E-02 | 4.59978E-02 | 5.40423E-02 |
| 2.0000 | 1.22148E-04 | 7.46070E-04 | 2.51545E-03 | 8.29603E-03 | 1.85097E-02 | 2.87006E-02 | 3.52611E-02 |
| 2.5000 | 7.08999E-05 | 4.30705E-04 | 1.36664E-03 | 4.30894E-03 | 1.05674E-02 | 1.77707E-02 | 2.27811E-02 |
| 3.0000 | 4.17638E-05 | 2.52855E-04 | 7.81141E-04 | 2.32154E-03 | 6.05482E-03 | 1.09491E-02 | 1.46169E-02 |
| 3.5000 | 2.48466E-05 | 1.50088E-04 | 4.57780E-04 | 1.29049E-03 | 3.48619E-03 | 6.72381E-03 | 9.33044E-03 |
| 4.0000 | 1.48878E-05 | 8.97824E-05 | 2.71949E-04 | 7.35541E-04 | 2.01838E-03 | 4.11977E-03 | 5.93218E-03 |
| 4.5000 | 8.96842E-06 | 5.40177E-05 | 1.62914E-04 | 4.27372E-04 | 1.17535E-03 | 2.52035E-03 | 3.75956E-03 |
| 5.0000 | 5.42491E-06 | 3.26433E-05 | 9.81567E-05 | 2.51905E-04 | 6.88408E-04 | 1.54028E-03 | 2.37641E-03 |
| 5.5000 | 3.29218E-06 | 1.97949E-05 | 5.93909E-05 | 1.50058E-04 | 4.05466E-04 | 9.40701E-04 | 1.49884E-03 |
| 6.0000 | 2.00314E-06 | 1.20369E-05 | 3.60525E-05 | 9.00840E-05 | 2.40082E-04 | 5.74293E-04 | 9.43578E-04 |
| 6.5000 | 1.22140E-06 | 7.33567E-06 | 2.19416E-05 | 5.43893E-05 | 1.42852E-04 | 3.50533E-04 | 5.93051E-04 |
| 7.0000 | 7.45996E-07 | 4.47853E-06 | 1.33807E-05 | 3.29760E-05 | 8.53744E-05 | 2.13934E-04 | 3.72185E-04 |
| 7.5000 | 4.56188E-07 | 2.73772E-06 | 8.17201E-06 | 2.00527E-05 | 5.12175E-05 | 1.30545E-04 | 2.33220E-04 |
| 8.0000 | 2.79093E-07 | 1.67441E-06 | 4.99419E-06 | 1.22149E-05 | 3.08141E-05 | 7.96128E-05 | 1.45858E-04 |
| 8.5000 | 1.70530E-07 | 1.02282E-06 | 3.04872E-06 | 7.43779E-06 | 1.85559E-05 | 4.84526E-05 | 9.09100E-05 |
| 9.0000 | 1.03549E-07 | 6.20941E-07 | 1.84979E-06 | 4.50390E-06 | 1.11273E-05 | 2.92899E-05 | 5.61994E-05 |
| 9.5000 | 6.14347E-08 | 3.68328E-07 | 1.09673E-06 | 2.66612E-06 | 6.53155E-06 | 1.72943E-05 | 3.38817E-05 |
| 10.0000 | 3.17138E-08 | 1.90106E-07 | 5.65825E-07 | 1.37372E-06 | 3.34080E-06 | 8.88277E-06 | 1.77477E-05 |

Table 2.5

Source Function for Albedo 0.9, Thickness 20

ALBEDO =0.9000 THICKNESS = 20.0000

| DEPTH | J(1) | J(2) | J(3) | J(4) | J(5) | J(6) | J(7) |
|---|---|---|---|---|---|---|---|
| 0. | 2.37857E-01 | 2.72191E-01 | 3.12443E-01 | 3.50108E-01 | 3.80502E-01 | 4.01766E-01 | 4.13530E-01 |
| 2.0000 | 2.19272E-03 | 1.35372E-02 | 4.01793E-02 | 9.02770E-02 | 1.57081E-01 | 2.18396E-01 | 2.57074E-01 |
| 4.0000 | 7.50041E-04 | 4.61642E-03 | 1.34726E-02 | 2.92557E-02 | 5.25648E-02 | 7.84377E-02 | 9.72742E-02 |
| 6.0000 | 2.61234E-04 | 1.60725E-03 | 4.68545E-03 | 1.01234E-02 | 1.81265E-02 | 2.74722E-02 | 3.47699E-02 |
| 8.0000 | 9.12544E-05 | 5.61401E-04 | 1.63628E-03 | 3.53285E-03 | 6.31466E-03 | 9.59848E-03 | 1.22408E-02 |
| 10.0000 | 3.18979E-05 | 1.96234E-04 | 5.71921E-04 | 1.23466E-03 | 2.20574E-03 | 3.35424E-03 | 4.28937E-03 |
| 12.0000 | 1.11510E-05 | 6.85995E-05 | 1.99930E-04 | 4.31594E-04 | 7.70948E-04 | 1.17240E-03 | 1.50071E-03 |
| 14.0000 | 3.89598E-06 | 2.39676E-05 | 6.98523E-05 | 1.50790E-04 | 2.69345E-04 | 4.09589E-04 | 5.24468E-04 |
| 16.0000 | 1.35434E-06 | 8.33177E-06 | 2.42825E-05 | 5.24184E-05 | 9.36301E-05 | 1.42380E-04 | 1.82336E-04 |
| 18.0000 | 4.50862E-07 | 2.77365E-06 | 8.08365E-06 | 1.74501E-05 | 3.11694E-05 | 4.73979E-05 | 6.07016E-05 |
| 20.0000 | 7.67640E-08 | 4.72243E-07 | 1.37632E-06 | 2.97106E-06 | 5.30691E-06 | 8.06996E-06 | 1.03352E-05 |

Table 2.6

Source functions for Albedo 1.0, Thickness 30

ALBEDO =1.0000     THICKNESS = 30.0000

| DEPTH | J(1) | J(2) | J(3) | J(4) | J(5) | J(6) | J(7) |
|---|---|---|---|---|---|---|---|
| 0. | 2.68771E-01 | 3.25769E-01 | 4.05381E-01 | 4.95188E-01 | 5.81497E-01 | 6.51106E-01 | 6.93493E-01 |
| 2.0000 | 1.08785E-02 | 6.72467E-02 | 1.94240E-01 | 4.05662E-01 | 6.69237E-01 | 9.13453E-01 | 1.07229E 00 |
| 4.0000 | 1.00808E-02 | 6.22639E-02 | 1.79090E-01 | 3.70802E-01 | 6.16470E-01 | 8.58267E-01 | 1.02368E 00 |
| 6.0000 | 9.32358E-03 | 5.75847E-02 | 1.65612E-01 | 3.42730E-01 | 5.69587E-01 | 7.94299E-01 | 9.49691E-01 |
| 8.0000 | 8.56873E-03 | 5.29224E-02 | 1.52202E-01 | 3.14970E-01 | 5.23417E-01 | 7.30000E-01 | 8.73116E-01 |
| 10.0000 | 7.81408E-03 | 4.82615E-02 | 1.38798E-01 | 2.87230E-01 | 4.77315E-01 | 6.65706E-01 | 7.96256E-01 |
| 12.0000 | 7.05946E-03 | 4.36008E-02 | 1.25394E-01 | 2.59491E-01 | 4.31219E-01 | 6.01417E-01 | 7.19363E-01 |
| 14.0000 | 6.30484E-03 | 3.89401E-02 | 1.11990E-01 | 2.31753E-01 | 3.85124E-01 | 5.37129E-01 | 6.42468E-01 |
| 16.0000 | 5.55023E-03 | 3.42795E-02 | 9.85859E-02 | 2.04015E-01 | 3.39029E-01 | 4.72841E-01 | 5.65573E-01 |
| 18.0000 | 4.79562E-03 | 2.96188E-02 | 8.51822E-02 | 1.76277E-01 | 2.92935E-01 | 4.08554E-01 | 4.88678E-01 |
| 20.0000 | 4.04102E-03 | 2.49583E-02 | 7.17786E-02 | 1.48540E-01 | 2.46841E-01 | 3.44267E-01 | 4.11783E-01 |
| 22.0000 | 3.28642E-03 | 2.02977E-02 | 5.83751E-02 | 1.20802E-01 | 2.00747E-01 | 2.79980E-01 | 3.34889E-01 |
| 24.0000 | 2.53182E-03 | 1.56371E-02 | 4.49715E-02 | 9.30646E-02 | 1.54653E-01 | 2.15693E-01 | 2.57994E-01 |
| 26.0000 | 1.77717E-03 | 1.09762E-02 | 3.15669E-02 | 6.53250E-02 | 1.08556E-01 | 1.51402E-01 | 1.81095E-01 |
| 28.0000 | 1.02169E-03 | 6.31016E-03 | 1.81477E-02 | 3.75551E-02 | 6.24085E-02 | 8.70405E-02 | 1.04111E-01 |
| 30.0000 | 2.17832E-04 | 1.34538E-03 | 3.86923E-03 | 8.00705E-03 | 1.30060E-02 | 1.85577E-02 | 2.21972E-02 |

We outline the procedure used to obtain Table 2.3 for the source function when the albedo is 0.3 and the thickness is 5.   Assume we write a FORTRAN program to do the following.   Understand that  $i = 1,2,...,7$.

Preliminaries

- Set  $\lambda = 0.3$, $N = 7$, $h = .005$.

- Read in the values of  $z_i$   and  $w_i$   given in Table 2.1 .

- Set  $x_1 = 0.5$, $x_2 = 1.0,...,x_{10} = 5$.
- Set  $t_1 = 0.5$, $t_2 = 1.0,...,t_{10} = 5$.

Initial Step at  $x = 0$

- Set  2N initial conditions (I.C.):
   $X_i(0) = 1$,  $Y_i(0) = 1$ ,by Eqs. (2.57) and (2.58).

- Integrate  2N differential equations (D.E.) for  $X_i(x)$, $Y_i(x)$,          Eqs. (2.54) and (2.55) for  $t = h$, $2h,...,0.5$.

First Print-out at  $x = 0.5$

- Print  $J_i(0.5,0.5) = (\lambda/4) X_i(0.5)$,
        $J_i(0.0,0.5) = (\lambda/4) Y_i(0.5)$.

- Set  N  new I.C. at  $x = 0.5$,
        $J_i(0.5,x) = (\lambda/4) X_i(x)$   by Eq. (2.56) .

- Adjoin N  new D.E. for  $J_i(0.5,x)$, Eq. (2.54).

- Integrate  3N differential equations for  $t = 0.5+h,...,1.0$.

Second Print-out at x = 1.0

- Print $J_i(1.0,1.0) = (\lambda/4)X_i(1.0)$,

  $J_i(0.5,1.0) = J_i(0.5,1.0)$,

  $J_i(0.0,1.0) = (\lambda/4) Y_i(1.0)$.

- Set N new I.C. at x = 1.0,

  $J_i(1.0,x) = (\lambda/4) X_i(x)$ by Eq. (2.56).

- Adjoin N new D.E. for $J_i(1.0,x)$, Eq. (2.54).

- Integrate 4N differential equations for
  t = 1.0+h,...,1.5.

...

Ninth Print-out at x = 4.5

- Print $J_i(0,4.5)$, $J_i(0.5,4.5)$,...,$J_i(4.5,4.5)$.

- Set N new I.C. at x = 4.5,

  $J_i(4.5,x) = (\lambda/4)X_i(x)$, by Eq. (2.56).

- Adjoin N new D.E. for $J_i(4.5,x)$, Eq. (2.54).

- Integrate 11N differential equations for
  t = 4.5+h,...,5.0.

Tenth Print-out at x = 5

- Print $J_i(0,5)$, $J_i(0.5,5)$,...$J_i(5,5)$.

Table 2.3 is the tenth-print-out. Note that we have produced nine other interesting tables along the way for slabs of smaller thicknesses and the same albedo. We found the equations to be exceptionally stable, and the solutions are well behaved. See the following graphs.

Selected graphs of the  X  and  Y  functions appear
in Figs. 2.1 - 2.6 for albedoes  0.4, 0.8  and  1.0.  We note
their monotonic behavior.  The limiting curves are labelled
"H", corresponding to semi-infinite thickness.

The source function for different values of the albedo
appears in Figs. 2.7-2.11.  In these graphs, one first notes
that three angles of incidence are represented.  Although not
apparent in Figs. 2.7  and 2.8, there are families of curves
for different slab thicknesses.  The horizontal axis is optical distance into the medium from the top, hence "optical
depth."  A curve extending to depth 8, for example, as in
Fig. 2.10, is the source function for a slab of thickness 8,
for the specified albedo and angle of incidence.  Five different slab thicknesses are represented in Fig.2.11.  The
numerical values used in plotting these graphs were obtained
by the initial value method.

When we plotted the graphs of the source functions
in Figs. 2.7 - 2.11, we first did a computer run and produced a table like one of those just presented.  We then
plotted the points, and drew smooth lines connecting the
points.  The graphs imply that we produced continuous functions of  t, while in reality we produced continuous functions of thickness.

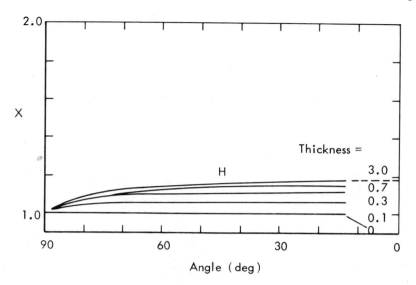

Figure 2.1   X   Functions for   $\lambda = 0.4$

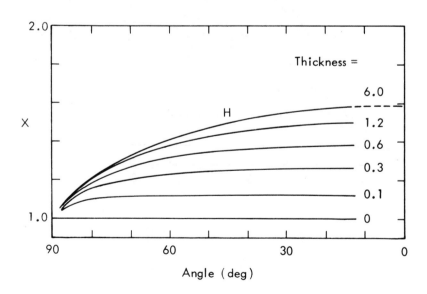

Figure 2.2   X   Functions for   $\lambda = 0.8$

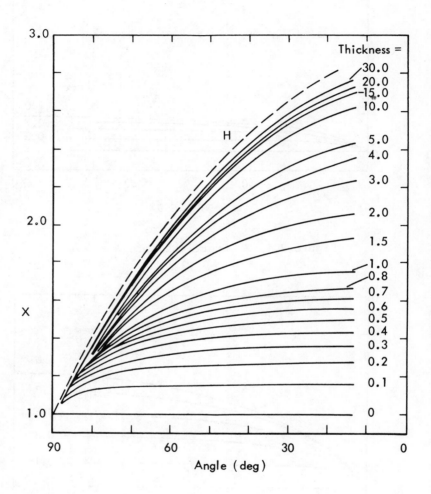

Figure 2.3   X   Functions for   $\lambda = 1.0$

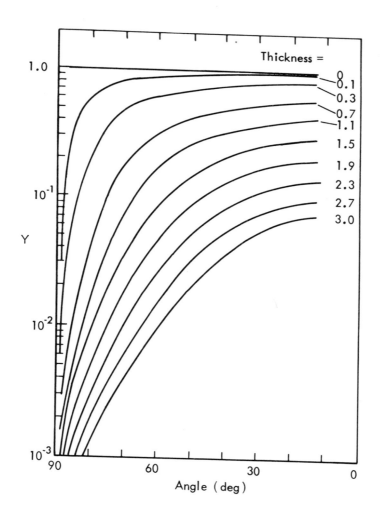

Figure 2.4  Y  Functions for  $\lambda = 0.4$

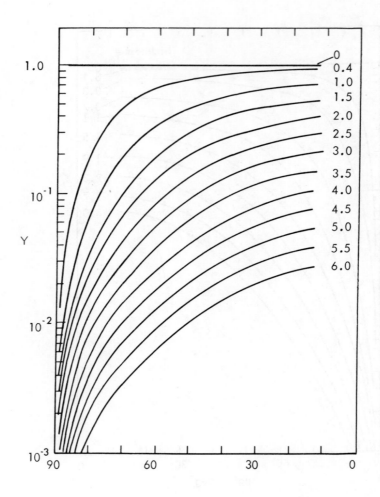

Figure 2.5   Y   Functions for   $\lambda = 0.8$

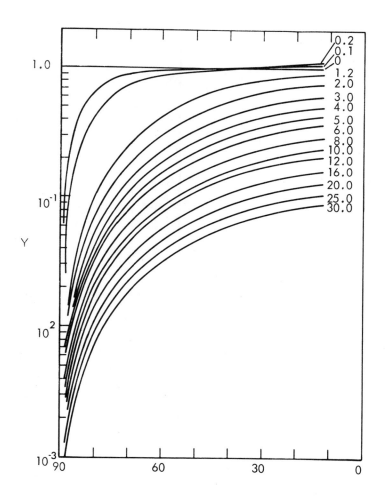

Figure 2.6  Y  Functions for  $\lambda = 1.0$

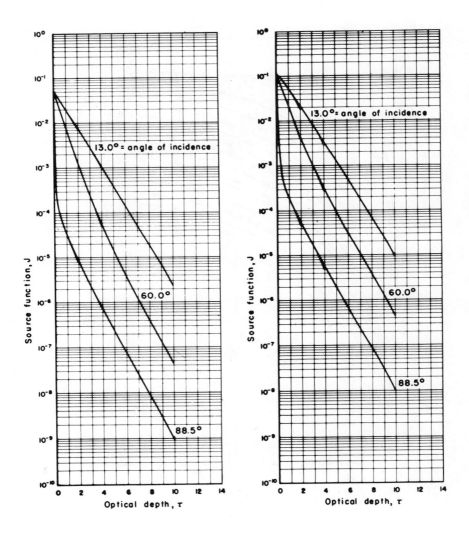

Figure 2.7  Several Source
            Functions for
            λ = 0.2

Figure 2.8  Several Source
            Functions for
            λ = 0.4

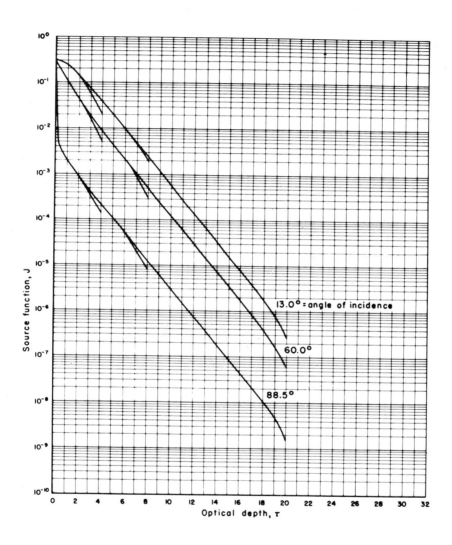

Figure 2.9   Several Source Functions for   $\lambda = 0.8$

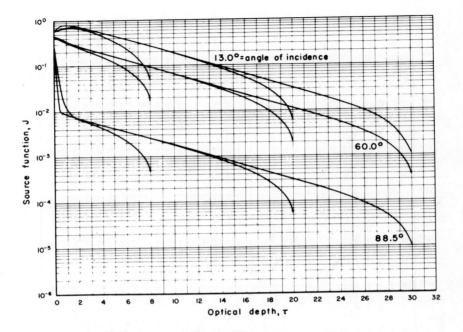

Figure 2.10   Several Source Functions for   $\lambda = 0.99$

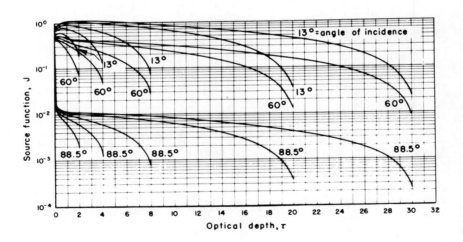

Figure 2.11   Several Source Functions for   $\lambda = 1.0$

## 2.6   MONODIRECTIONAL ILLUMINATION - INTERNAL INTENSITY FUNCTION

The internal intensity function  $I(t,v,x,u)$  satisfies the initial value problem

$$I_x(t,v,x,u) = u^{-1} I(t,v,x,u)$$

$$+ 2 J(x,x,u) \int_0^1 I(t,v,x,u') \, du'/u'$$

$$+ g(t,v,x,u) ,$$

$$t \leq x, \quad 0 \leq u \leq 1, \quad -1 \leq v \leq 1 , \qquad (2.60)$$

where

$$g(t,v,x,u) = \begin{cases} 0 , & 0 \leq v \leq 1 , \\ J(x,x,u)(-v)^{-1} e^{+(x-t)/v} , & \\ & -1 \leq v \leq 0 , \end{cases} \qquad (2.61)$$

with initial conditions

$$I(t,v,t,u) = \begin{cases} r(t,v,u) , & 0 \leq v \leq 1 , \\ 0 , & -1 \leq v \leq 0 , \end{cases} \qquad (2.62)$$

The reflection function  $r$  may be known through its own initial value problem, or may be expressed in terms of  $X$  and  $Y$  functions.  See Chapter 4.  This system is to be augmented by the Cauchy system for the source function  $J$.

A result from a computation for a slab of thickness
10 with $\lambda = 1$ is shown in Table 2.7. Here, the altitude
is t = 5. The numbers of the left indicate the propagating
angle. Each column corresponds to a single incident angle
and fourteen propagating directions. See Table 2.1 for exact
values of the angles. The leftmost angle corresponds to the
largest angle of incidence.

Internal intensity functions are shown in the next
group of graphs, Fig. 2.12 on pages 73 -78. In the
first pair of graphs, the albedo is $\lambda = 0.1$ and the thickness is x = 2. The incident angles are $13^o$ and $60^o$,
respectively. The remaining graphs are for $\lambda = 0.5, 0.9$
and 1.0 with x = 2, and for $\lambda = 1.0$ with x = 10.

Note again the distinct difference in the curves for
conservative scattering, $\lambda = 1$, in contrast with those for
$\lambda < 1$.

Figures 2.13 and 2.14 show the variation of intensity
with depth in the medium for $\lambda = 0.9$ and 1.0. The incident
angle is held constant. The intensity plotted is that for
a specific propagation direction, e.g. "$60^o$ up" refers to the
angle whose cosine is v = 0.5. Note the exponential drop-
off when the albedo is less than unity(Fig. 2.13) and the
linearity of the curves when scattering is conservative
(Fig. 2.14).

The above-mentioned figures were drawn from numerical
results obtained by the initial value method. In contrast,
the unbounded behavior of the curves in Fig. 2.15 was the
result of a direct integration of the transfer equation,
which is extremely unstable.

Table 2.7   Internal Intensities for a Slab of Thickness 10 with Albedo 1.0 and at Altitude 5.

| | | | | | | | | |
|---|---|---|---|---|---|---|---|---|
| 1 | U | 5.20202E-03 | 3.64529E-02 | 1.04842E-01 | 2.17002E-01 | 3.60732E-01 | 5.02727E-01 | 6.00286E-01 |
|   | V | 5.95607E-03 | 3.67868E-02 | 1.05802E-01 | 2.18993E-01 | 3.64047E-01 | 5.07323E-01 | 6.05728E-01 |
| 2 | U | 5.79416E-03 | 3.57867E-02 | 1.02925E-01 | 2.13031E-01 | 3.54121E-01 | 4.93553E-01 | 5.89410E-01 |
|   | V | 6.06296E-03 | 3.74471E-02 | 1.07702E-01 | 2.22934E-01 | 3.70609E-01 | 5.16404E-01 | 6.16459E-01 |
| 3 | U | 5.61980E-03 | 3.47097E-02 | 9.98272E-02 | 2.06614E-01 | 3.43442E-01 | 4.78715E-01 | 5.71785E-01 |
|   | V | 6.23742E-03 | 3.85249E-02 | 1.10805E-01 | 2.29378E-01 | 3.81340E-01 | 5.31193E-01 | 6.33854E-01 |
| 4 | U | 5.40900E-03 | 3.34076E-02 | 9.60819E-02 | 1.98858E-01 | 3.30541E-01 | 4.60768E-01 | 5.50430E-01 |
|   | V | 6.44930E-03 | 3.98350E-02 | 1.14584E-01 | 2.37251E-01 | 3.94374E-01 | 5.48906E-01 | 6.54454E-01 |
| 5 | U | 5.19812E-03 | 3.21051E-02 | 9.23356E-02 | 1.91102E-01 | 3.17642E-01 | 4.42811E-01 | 5.29036E-01 |
|   | V | 6.66219E-03 | 4.11512E-02 | 1.18376E-01 | 2.45068E-01 | 4.06970E-01 | 5.65507E-01 | 6.73361E-01 |
| 6 | U | 5.02414E-03 | 3.10306E-02 | 8.92448E-02 | 1.84703E-01 | 3.07002E-01 | 4.27994E-01 | 5.11370E-01 |
|   | V | 6.83131E-03 | 4.21907E-02 | 1.21328E-01 | 2.50977E-01 | 4.16096E-01 | 5.77047E-01 | 6.86148E-01 |
| 7 | U | 4.91727E-03 | 3.03705E-02 | 8.73464E-02 | 1.80774E-01 | 3.00468E-01 | 4.18892E-01 | 5.00512E-01 |
|   | V | 6.92746E-03 | 4.27765E-02 | 1.22962E-01 | 2.54146E-01 | 4.20792E-01 | 5.82748E-01 | 6.92285E-01 |

FIGURE 2.12

Internal Intensity Functions

on pages 73 - 78

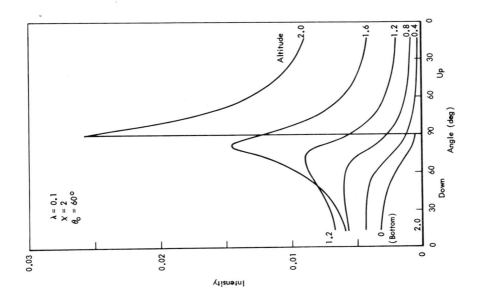

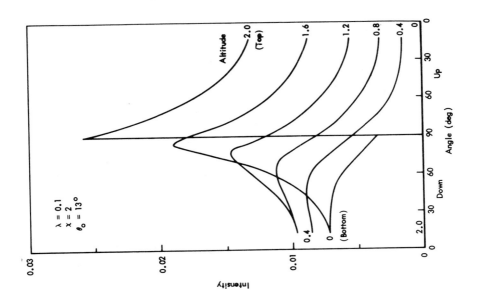

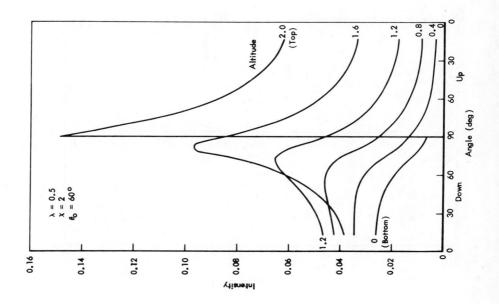

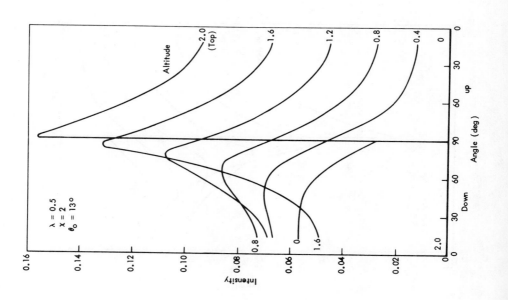

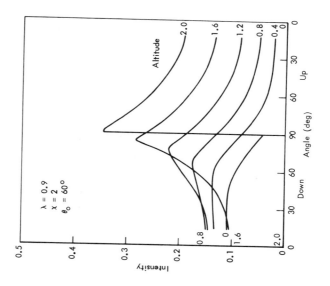

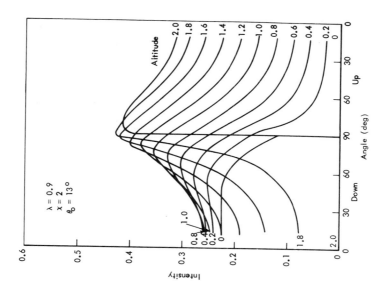

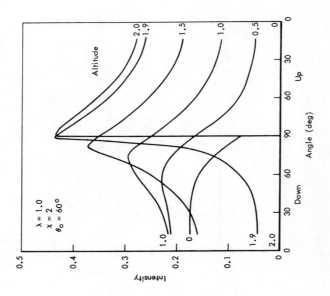

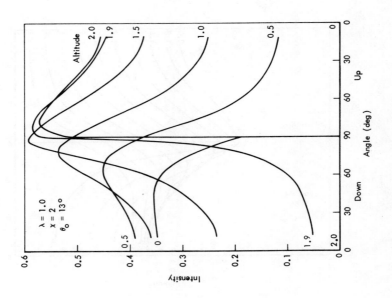

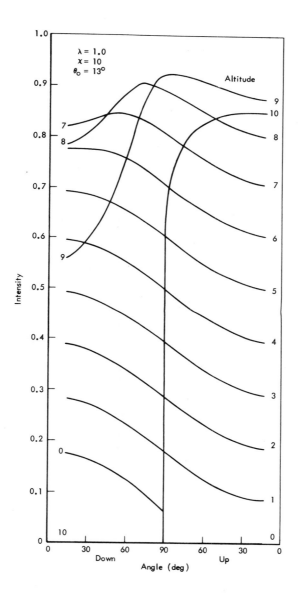

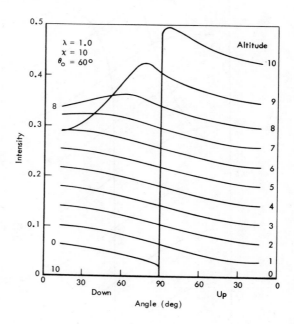

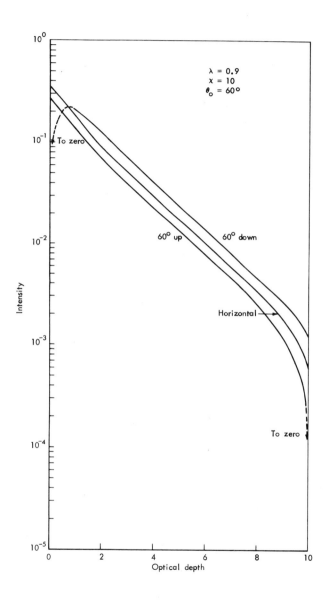

Figure 2.13   Intensity as a Function of Depth
              for Albedo 0.9

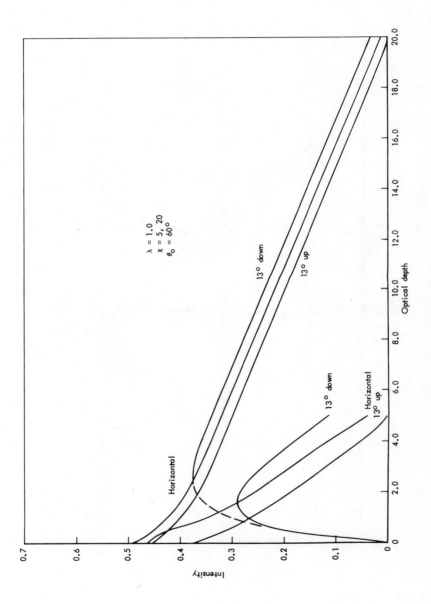

Figure 2.14 Intensity as a Function of Depth for Albedo 1.0

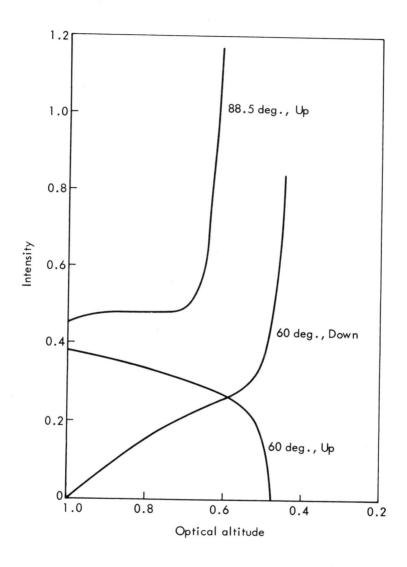

Figure 2.15   Intensities Produced by Integrating Unstable
              Transfer Equation

2.7    MONODIRECTIONAL ILLUMINATION - REFLECTED AND TRANS-
       MITTED INTENSITIES

       Reflection and transmission functions are also solu-
tions of Cauchy systems (see Chapter 6). Reflected and
transmitted intensities are shown in Figs. 2.16-2.25 for
albedoes 0.2, 0.5, 0.8, 0.95 and 1.0, and for a variety of
thicknesses. There are three different angles of incidence,
88.5, 45.3, and 13.0 degrees. Each curve represents a
fixed thickness.

       In the case of reflected intensities, the largest
thickness indicates that at that thickness, the medium is
semi-infinite in effect. The only exception is the conservative case, $\lambda = 1$, at angles of incidence 45.3 and
13.0 degrees. The dashed curves represent the semi-infinite
reflection functions.

       Transmitted intensities drop off approximately ex-
ponentially for large thicknesses.

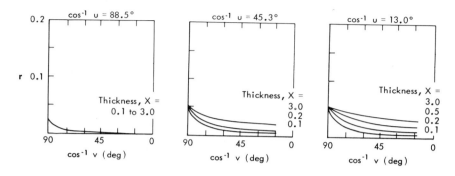

Figure 2.16   Reflected  intensities   r   for albedo   $\lambda$ = 0.2

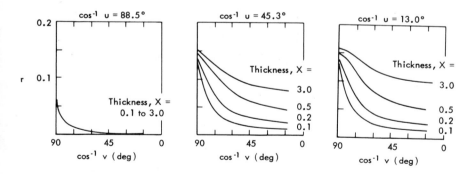

Figure 2.17   Reflected  intensities   r   for albedo   $\lambda$ = 0.5

84                                          MULTIPLE SCATTERING

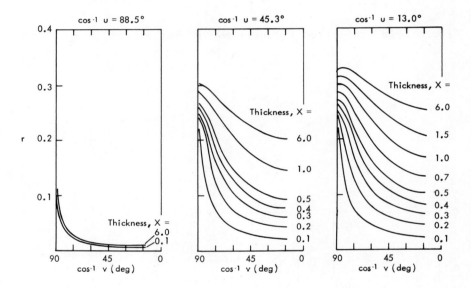

Figure 2.18   Reflected Intensities   r   for Albedo   $\lambda = 0.8$

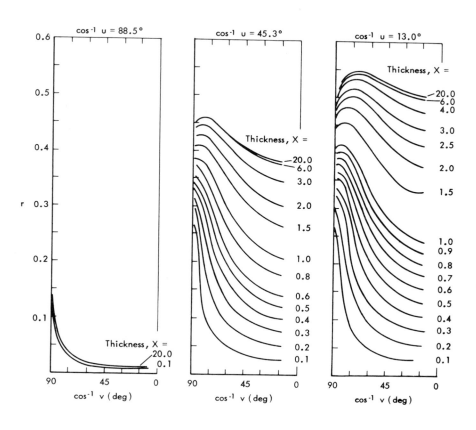

Figure 2.19   Reflected Intensities   r   for Albedo   $\lambda = 0.95$

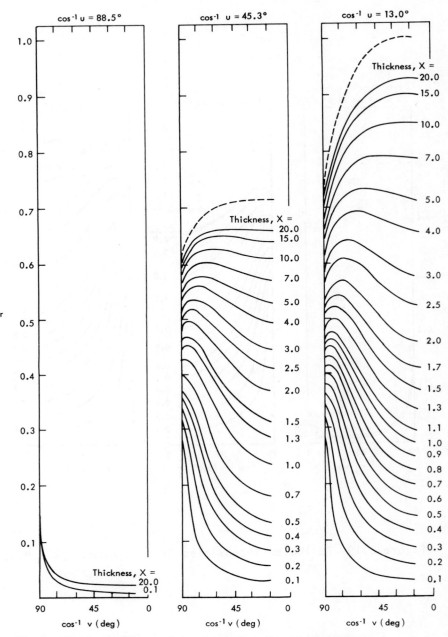

Figure 2.20  Reflected Intensities  r  for Albedo  $\lambda = 1.0$

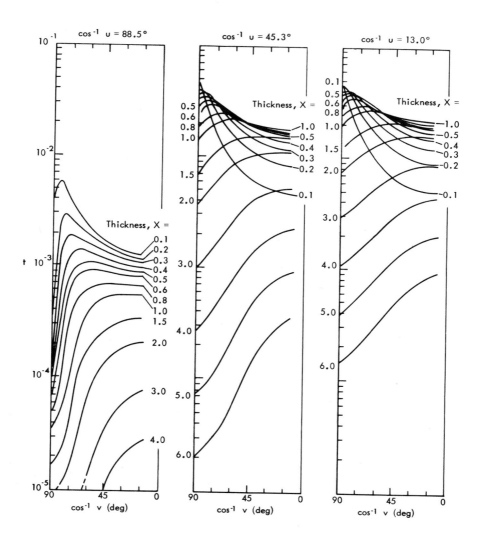

Figure 2.21   Transmitted Intensities   t   for Albedo   $\lambda$ = 0.2

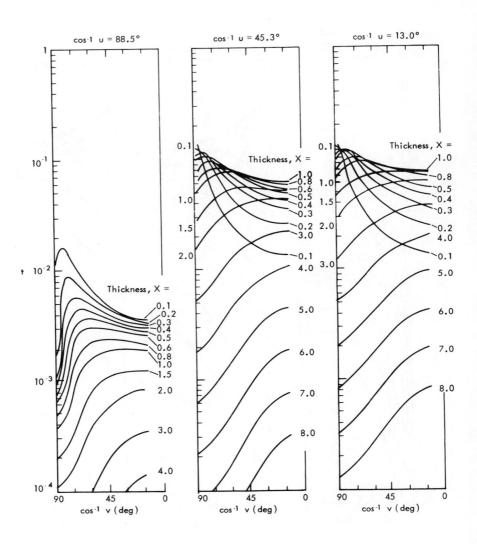

Figure 2.22   Transmitted Intensities   t   for Albedo   $\lambda = 0.5$

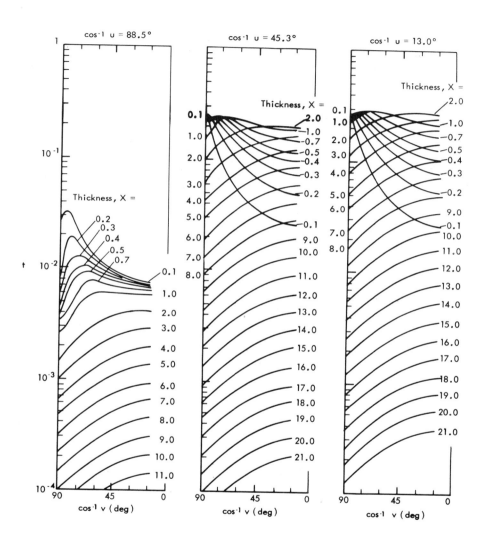

Figure 2.23   Transmitted Intensities   t   for Albedo   $\lambda$ = 0.8

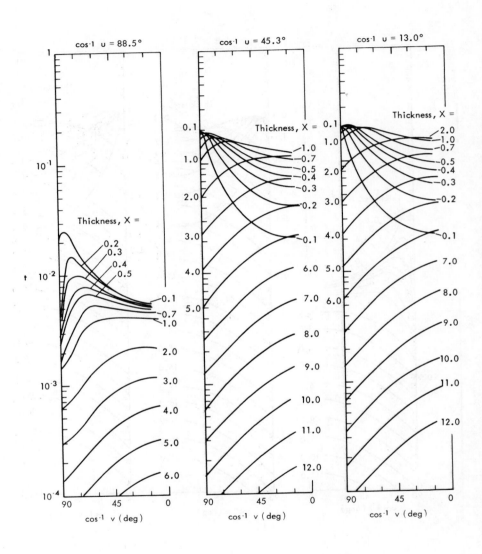

Figure 2.24   Transmitted Intensities   t   for Albedo   $\lambda = 0.95$

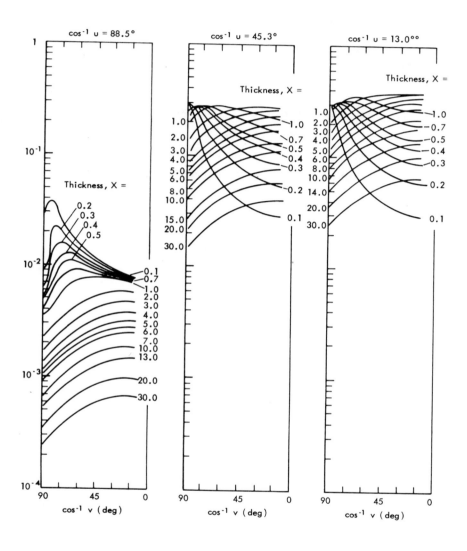

Figure 2.25  Transmitted Intensities  t  for Albedo  $\lambda = 1.0$

2.8    DISCUSSION

In this chapter we have introduced the basic source and intensity functions and have shown that they satisfy Cauchy systems. These initial value problems are reduced to ordinary differential equations by using the method of lines. Then results of numerical experiments are given in the form of tables and graphs.

In the next chapter we turn to a new approach to the auxiliary problem based upon isotropic illumination.

## EXERCISES

1.  Derive a Cauchy system for the emergence function $p$ .

2.  Derive a Cauchy system for the reflection and transmission functions.

3.  Derive a Cauchy system for the internal intensity function $I$. Write out the system of approximating ordinary differential equations.

4.  Write computer programs and calculate the functions discussed in this chapter.

## REFERENCES

1.  R. Bellman, H. Kagiwada, R. Kalaba, "Invariant Imbedding and a Reformulation of the Internal Intensity Problem in Radiative Transfer Theory," *Monthly Notices of the Royal Astronomical Society*, v. 132, 1966, pp. 183-191.

2.  R. Bellman, H. Kagiwada, R. Kalaba, and M. Prestrud, *Invariant Imbedding and Time Dependent Transport Processes*, American Elsevier Publishing Co., New York, 1964.

# CHAPTER 3

## THE BASIC PROBLEM: b AND h FUNCTIONS

### 3.1   OMNIDIRECTIONAL ILLUMINATION

The functions  b  and  h  of radiative transfer are
internal intensities due to omnidirectional sources at the
boundaries and as such are of interest in themselves.  But
more remarkable is the fact that they also yield, via simple
algebraic formulas, all of the quantities of interest for
the case of monodirectional illumination, the auxiliary
problem which was discussed in Chapter 2.

We consider the basic problem for a horizontal,
plane-parallel slab of finite optical thickness  x.  The
slab is homogeneous, with albedo for single scattering  $\lambda$,
and it scatters energy isotropically.  Let there be omnidirectional sources of energy at the top of the slab which
emit in all downward directions with the strength of one
unit of energy per unit of horizontal area per unit of solid
angle per unit of time.  This input is referred to as unit
omnidirectional sources at the top.  Let  $b = b(t,v,x)$  be
defined as

$b(t,v,x)$ = the total intensity at the altitude  t
in a direction whose direction cosine with respect
to the upward vertical is  v  $(-1 \leq v \leq +1)$,  the slab
thickness being  x,  and due to unit omnidirectional
sources at the top.                              (3.1)

Similarly, let omnidirectional sources of energy be incident
at the bottom in all upward directions and having the strength
of one unit of energy per unit of horizontal area per unit
of solid angle per unit time.  This is referred to as unit
omnidirectional sources at the bottom.  The function
$h = h(t,v,x)$  is defined as

$h(t,v,x)$ = the total intensity at the altitude  t
in a direction whose direction cosine with respect
to the upward vertical is  v  $(-1 \leq v \leq +1)$,  the slab
thickness being  x,  and due to unit omnidirectional
sources at the bottom.                           (3.2)

In addition, we define the source function for the
case of omnidirectional sources at the top,

$\Phi(t,x)$ = the total rate of production of scattered
energy per unit volume per unit solid angle at the
altitude  t,  the slab thickness being  x,  and due
to unit omnidirectional sources at the top.  (3.3)

Some source functions  $\Phi(t,x)$  are presented in
Fig. 3.1 for slabs with albedos 0.5, 0.9 and 1.0, and thicknesses 2, 5 and 10. This source function is infinite at the top.

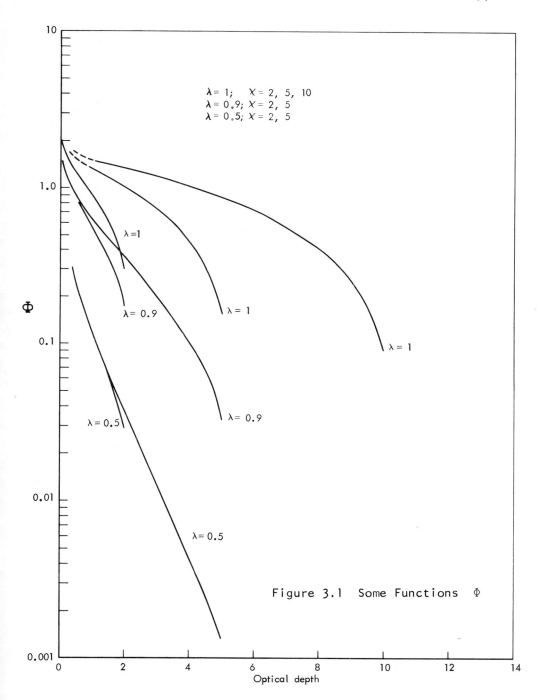

Figure 3.1  Some Functions  Φ

The intensity function  b  satisfies the differential equation

$$v \, \frac{\partial b(t,v,z)}{\partial t} = - b(t,v,x) + \Phi(t,x) , \qquad (3.4)$$

and the boundary conditions

$$b(x,v,x) = - \frac{1}{v} , \qquad\qquad -1<v<0 , \quad (3.5)$$

$$b(0,v,x) = 0 , \qquad\qquad 0<v<1 . \quad (3.6)$$

Solving for  b  in terms of  $\Phi$ , we obtain

$$b(t,v,x) = \begin{cases} v^{-1} \int_0^t e^{-(t-y)/v} \Phi(y,x) \, dy, & 0<v<1 , \\[2mm] - v^{-1} \left[ e^{-(t-x)/v} + \int_t^x e^{-(t-y)/v} \, \Phi(y,x) \, dy \right], & \\ & \qquad -1<v<0 . \quad (3.7) \end{cases}$$

This last Eq. (3.7), can, of course, be interpreted physically.

The source function  $\Phi$  is related to the intensity function  b  through the formula

$$\Phi(t,x) = \frac{\lambda}{2} \int_{-1}^{+1} b(t,v,x) \, dv , \qquad (3.8)$$

which expresses the rate of scattering as due to the scattering of the beams of radiation of total intensity  b. There is no additional term since all energy flowing at a point is accounted for in the function  b.

Elimination of  b  from the last two equations produces the linear integral equation for the source function
$\Phi(t,x)$ ,

$$\Phi(t,x) = \frac{\lambda}{2} E_1(x-t) + \frac{\lambda}{2} \int_0^x E_1(|t-y|)\Phi(y,x)\ dy\ .$$

$$(3.9)$$

We can physically understand the forcing term directly as
the rate of scattering, in a unit of volume centered at  t,
of the reduced incident radiation coming in from all directions,

$$\frac{\lambda}{2} E_1(x-t) = \int_0^1 1 \cdot e^{-(x-t)/z} \cdot \frac{1}{z} \cdot \frac{\lambda}{4\pi} \cdot 2\pi dz$$

$$= \frac{\lambda}{2} \int_0^1 e^{-(x-t)/z} \frac{dz}{z}\ . \qquad (3.10)$$

The integral term is, of course, the contribution due to
the multiply scattered radiation.

Regarding the intensity function  $h(t,v,x)$  due to
omnidirectional illumination at the bottom, the equations
are

$$v\ \frac{\partial h(t,v,x)}{\partial t} = -h(t,v,x) + \Phi(x-t,x)\ , \qquad (3.11)$$

$$h(x,v,x) = 0\ , \qquad\qquad -1<v<0\ , \quad (3.12)$$

$$h(0,v,x) = \frac{1}{v}\ , \qquad\qquad 0<v<1\ , \quad (3.13)$$

and

$$
h(t,v,x) = \begin{cases} v^{-1}\left[e^{-t/v} + \int_0^t e^{-(t-y)/v}\Phi(x-y,x)\,dy\right], \\ \qquad\qquad\qquad\qquad\qquad <v< \quad, \\ \\ -\,v^{-1}\int_t^x e^{-(t-y)/v}\Phi(x-y,x)\,dy \;, \\ \qquad\qquad\qquad\qquad\qquad -1<v<0 \;, \quad (3.14) \end{cases}
$$

It is easy to observe that the  b  and  h  functions are really one, since

$$
h(t,v,x) = b(x-t,-v,x) \; . \tag{3.15}
$$

## 3.2    RELATIONSHIPS BETWEEN BASIC AND AUXILIARY PROBLEMS

By comparison of the integral equations for the source functions  J  and  $\Phi$,  it is apparent that  $\Phi$  is readily expressed in terms of  J  as

$$
\Phi(t,x) = 2\int_0^1 J(t,x,z)\,\frac{dz}{z} \; . \tag{3.16}
$$

From this equation, together with the formulas for the internal intensities  b  and  I  in terms of their respective source functions, we have

$$
b(t,v,x) = \begin{cases} 2\int_0^1 I(t,v,x,z)\,\dfrac{dz}{z} \;, & 0<v<1 \\ \\ 2\int_0^1 I(t,v,x,z)\,\dfrac{dz}{z} - v^{-1}e^{-(t-x)/v} \;, \\ & -1<v<0 \; . \quad (3.17) \end{cases}
$$

These are the obvious interconnections between the monodirectional and omnidirectional problems. Let us now turn
to the more subtle, and more remarkable, relations.

First, we derive a decomposition formula which
expresses the internal intensity function  I  in terms of
the  X  and  Y  functions, the source function  J,  and the
b  and  h  functions. The point of this decomposition is
that the internal intensity function is expressed in terms
of functions of fewer variables.

We first consider the case of monodirectional illumination of the slab. Recall that  $I(t,v,x,z)$  denotes
the intensity at the altitude  t  in a direction whose
cosine is  v,  the slab thickness being  x  and the incident
direction cosine being  z.  In addition, recall that the
source function is denoted by  $J(t,x,z)$.  Now consider the
change that takes place in the intensity when we move from
altitude  t  to altitude  $t + \Delta$.  Following a particular
beam, we find

$$I(t+\Delta,v,x,z) = I(t,v,x,z) - \Delta v^{-1} I(t,v,x,z)$$

$$+ \Delta v^{-1} J(t,x,z) + o(\Delta) \ . \qquad (3.18)$$

In the limit as  $\Delta$  tends to 0, this becomes the transport
equation,

$$v \ I_t = -I + J \ . \qquad (3.19)$$

We can also achieve the same increment in altitude
by removing a slab of thickness  $\Delta$  from the top and adding
it to the bottom. This leads to the equation

$$I(t+\Delta,v,x,z) - I(t,v,x,z) = \Delta z^{-1} I(t,v,x,z)$$
$$- J(x,x,z) \Delta b(t,v,x) + J(0,x,z) \Delta h(t,v,x)$$
$$+ o(\Delta) . \qquad (3.20)$$

The first two terms on the right-hand side of Eq. (3.20) represent the change in intensity due to removal of the top layer, and the third represents the increment due to the addition of the bottom layer. The limiting form of Eq. (3.20) as $\Delta$ tends to 0 is

$$I_t(t,v,x,z) = z^{-1}I(t,v,x,z) - J(x,x,z) b(t,v,x)$$
$$+ J(0,x,z) h(t,v,x) . \qquad (3.21)$$

By eliminating the partial derivative $I_t$ from Eqs. (3.19) and (3.21), we obtain the formula

$$(z^{-1}+v^{-1}) I(t,v,x,z) = v^{-1}J(t,x,z) + J(x,x,z) b(t,v,x)$$
$$- J(0,x,z) h(t,v,x) . \qquad (3.22)$$

Upon introducing the X and Y functions,

$$J(x,x,z) = (\lambda/4) X(x,z) ,$$
$$J(0,x,z) = (\lambda/4) Y(x,z) , \qquad (3.23)$$

we find the desired decomposition formula for $I(t,v,x,z)$ ,

$$(z^{-1}+v^{-1}) I(t,v,x,z) = v^{-1}J(t,x,z)$$
$$+ (\lambda/4) X(x,z) b(t,v,x) -(\lambda/4) Y(x,z) h(t,v,x).$$
$$(3.24)$$

An important point to notice is that $I = I(t,v,x,z)$, a function of four variables, is expressed in terms of functions of fewer variables.

It is next shown that the functions X, Y, J and $\Phi$ may be expressed in terms of the functions b and h.

From the physical meanings of the X and Y functions as emergence probabilities, it follows that

$$X(x,z) = 1 + zb(x,z,x) , \qquad (3.25)$$

$$Y(x,z) = zh(x,z,x) . \qquad (3.26)$$

Let us establish Eq. (3.25) in greater detail. The probability that a particle absorbed at the top will ultimately emerge from the top with a direction cosine between v and $v + dv$ and azimuth between $\phi$ and $\phi + d\phi$ is $p(x,x,v)dvd\phi$. It may be expressed in the form

$$p(x,x,v) \, dvd\phi = \frac{\lambda}{4\pi} \, dvd\phi$$

$$+ \frac{\lambda}{4\pi} \cdot v \, b(x,v,x) \, dvd\phi . \qquad (3.27)$$

The first term on the right hand side is the probability of emission directly into the desired angular interval. The second term is the probability of emission downwards into the medium where the particle is multiply scattered and emerges in the proper interval. It follows that

$$p(x,x,v) = \frac{\lambda}{4\pi} [1 + vb(x,v,x)] . \qquad (3.28)$$

But

$$p(x,x,v) = \frac{1}{\pi} J(x,x,v)$$

$$= \frac{\lambda}{4\pi} X(x,v) \ . \tag{3.29}$$

Therefore, a formula for  X  is

$$X(x,v) = 1 + vb(x,v,x) \ . \tag{3.30}$$

The equation for  Y  is established in a similar manner.

From the transport equation for  b, it follows that

$$\Phi(t,x) = b(t,0,x) \ , \tag{3.31}$$

provided that

$$\lim_{v \to 0} vb_t(t,v,x) = 0 \ . \tag{3.32}$$

The expression for  J  in terms of  b  and  h  may be obtained in any of several ways. Perhaps the simplest is to put  v = - z  into the decomposition formula. This leads to the relations

$$J(t,x,z) = (\lambda/4) \ X(x,z) \ b(t,-z,x)z$$
$$- (\lambda/4) \ Y(x,z) \ h(t,-z,x)z \ , \tag{3.33}$$

$$J(t,x,z) = (\lambda/4) \ [1 + zb(x,z,x)] \ b(t,-z,x)z$$
$$- (\lambda/4) \ zh(x,z,x) \ h(t,-z,x)z \ . \tag{3.34}$$

The formula for  I  becomes

$$I(t,v,x,z) = \frac{\lambda}{4} \frac{z}{z+v} \{X(x,z)[vb(t,v,x) + zb(t,-z,x)]$$

$$- Y(x,z)[vh(t,v,x) + zh(t,-z,x)]\},$$

$$(3.35)$$

or

$$I(t,v,x,z) = \frac{\lambda}{4} \frac{z}{z+v} \{[1 + zb(x,z,x)][vb(t,v,x)$$

$$+ zb(t,-z,x)] - [zh(x,z,x)] \cdot$$

$$[vh(t,v,x) + zh(t,-z,x)]\}. \qquad (3.36)$$

To illustrate the use of the above relations, let us calculate X, Y, J, and I with the aid of the tables of b and h in Appendix C. Consider the case for which $\lambda = 1$, $x = 1$, $t = 0.5$, $v = \cos 60° = 0.5$, $z = \cos 60° = 0.5$. The Appendix yields the values

$$b(t,v,x) = 0.47436 ,$$

$$h(t,v,x) = 1.5721 ,$$

$$b(x,u,x) = 1.1481 ,$$

$$h(x,u,x) = 1.0001 ,$$

$$b(t,-u,x) = 1.5721 ,$$

$$h(t,-u,x) = 0.47436 . \qquad (3.37)$$

Then by the foregoing equations, the desired quantities are found to be

$$X(x,z) = 1.57405 \, ,$$

$$Y(x,z) = 0.50005 \, ,$$

$$J(t,x,z) = 0.279067 \, ,$$

$$I(t,v,x,z) = 0.13737 \, . \tag{3.38}$$

Independent calculations confirm these values of X, Y, J, and I .

The emergent intensity at the top is denoted $r(v,z,x)$ , defined as

$$r(v,z,x) = I(x,v,x,z) \, . \tag{3.39}$$

This reflected intensity is expressed in terms of a function $R(v,z,x)$ ,

$$r(v,z,x) = R(v,z,x)/4v \, . \tag{3.40}$$

Similarly for the transmitted intensity $t(v,z,x)$ ,

$$t(v,z,x) = I(0,-v,x,z) \tag{3.41}$$

and the transmission function $T(v,z,x)$ ,

$$t(v,z,x) = T(v,z,x)/4v \, . \tag{3.42}$$

Equations for the functions R and T are

$$\left(\frac{1}{z} + \frac{1}{v}\right) \frac{R(v,z,x)}{4v} = \frac{1}{v}\frac{\lambda}{4} X(x,z) + \frac{\lambda}{4} X(x,z) \, b(x,v,x)$$
$$- \frac{\lambda}{4} Y(x,z) \, h(x,v,x) \, , \tag{3.43}$$

$$\left(\frac{1}{z} + \frac{1}{v}\right) R(v,z,x) = \lambda[X(x,z) \ X(x,v) - Y(x,z) \ Y(x,v)],$$
$$(3.44)$$

$$\left(\frac{1}{z} + \frac{1}{v}\right) R(v,z,x) = \lambda\{[1 + zb(x,z,x)][1 + vb(x,v,x)]$$
$$- zh(x,z,v) \ vh(x,v,x)\} , \qquad (3.45)$$

and

$$\left(\frac{1}{z} - \frac{1}{v}\right) \frac{T(v,z,x)}{4v} = - \frac{1}{v} \frac{\lambda}{4} Y(x,z) + \frac{\lambda}{4} X(x,z) \ h(x,v,x)$$
$$- \frac{\lambda}{4} Y(x,z) \ b(x,v,x) , \qquad (3.46)$$

$$\left(\frac{1}{z} - \frac{1}{v}\right) T(v,z,x) = \lambda[X(x,z) \ Y(x,v) - Y(x,z) \ X(x,v)],$$
$$(3.47)$$

$$\left(\frac{1}{z} - \frac{1}{v}\right) T(v,z,x) = \lambda\{[1 + zb(x,z,x)] \ vh(x,v,x)$$
$$- zh(x,z,x) \ [1 + vb(x,v,x)]\}.$$
$$(3.48)$$

It should be noted that the method for determining internal intensity I involves no matrix inversions, and overcomes a serious difficulty associated with the straight-forward use of invariance principles for this purpose. The relations which are here obtained show how to express the functions X, Y, Φ, J, I,R and T algebraically in terms of the basic functions b and h. Such relations are of interest in studying the analytic properties of the func-tions X, Y, ..., T. Of equal interest is their computational

significance.  Extensive tables of the functions  b  and
h  have been computed.  Some of these are given in Appendix
C.  These can be used, together with the given formulas and
a desk calculator, to compute the standard functions.  Recall that the two functions  b  and  h  are actually one.
For a given atmospheric thickness and albedo for single
scattering it is no more difficult to tabulate these functions than the source function.  The advantages in tabulating
b  and  h, though, are manifest.  The FORTRAN program for
computing  b  and  h  is listed in Appendix B.  In several
seconds, or at most several minutes, any required values
of  X, Y, ..., T  can be calculated for a wide range of
values of  x  and  $\lambda$.


3.3      CAUCHY SYSTEM FOR  b  AND  h  FUNCTIONS


          Let us physically derive the differential equation
for  h(t,v,x)  in which the thickness,  x, is the independent
variable.  Consider the internal intensity at altitude  t
in a direction whose cosine is  v  (-1$\leq$v$\leq$+1).  This is sketched
in the left half of Fig. 3.2.  Next, let a slab of thickness
x + $\Delta$  be formed by the addition of a layer of thickness  $\Delta$
to the top of the previous slab of thickness  x;  see the
right half of Fig. 3.2.  We obtain the relation

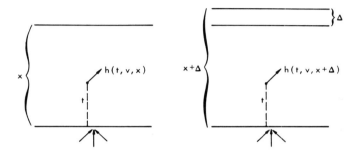

Figure 3.2   The Intensity  h   in Slabs of Thickness
             x   and   x + Δ

$$h(t,v,x+\Delta) = h(t,v,x) + \Delta\Phi(0,x)\ b(t,v,x) + o(\Delta)\ ,$$
$$(3.49)$$

where $\Phi(0,x)$ is the source function at altitude $x$
due to omnidirectional sources at the bottom. We can write

$$\Phi(0,x) = 2\int_0^1 J(0,x,z)\ \frac{dz}{z}\ ,\qquad\qquad (3.50)$$

but since

$$J(0,z,x) = \frac{\lambda}{4} Y(x,z)\ ,\qquad\qquad (3.51)$$

we have

$$\Phi(0,x) = \frac{\lambda}{2} \int_0^1 Y(x,z)\ \frac{dz}{z}\ .\qquad\qquad (3.52)$$

In the limit as $\Delta$ tends to 0, Eq. (3.49) becomes

$$h_x(t,v,x) = \frac{\lambda}{2} b(t,v,x) \int_0^1 Y(x,z)\ \frac{dz}{z}\ .\qquad (3.53)$$

The derivation of the equation for $b$ is slightly
more complex. Refer to Fig. 3.3. Add a slab of thickness
$\Delta$ to the bottom of the slab. This has two major effects:
additional sources are created at the bottom, and the
altitude is increased. We obtain the relation

$$b(t,v,x+\Delta) = b(t,v,x) + \Delta\Phi(0,x)\ h(t,v,x)$$

$$-\ \frac{\partial b(t,v,x)}{\partial t}\ \Delta + o(\Delta)\ .\qquad\qquad (3.54)$$

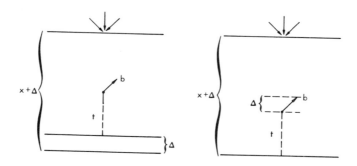

Figure 3.3   The Intensity  b   in Slabs of Thickness
             x   and   x + Δ

For  ∂b/∂t  we use the transport equation

$$v \, \frac{\partial b(t,v,x)}{\partial t} = - \, b(t,v,x) + \Phi(t,x) \; , \qquad (3.55)$$

so that Eq. (3.54) becomes

$$b(t,v,x+\Delta) = b(t,v,x) + \Delta\Phi(0,x) \, h(t,v,x)$$

$$+ \frac{\Delta}{v} \, [b(t,v,x) - \Phi(t,x)] + o(\Delta) \; . \quad (3.56)$$

The limiting form of this equation is

$$b_x(t,v,x) = \frac{\lambda}{2} \, h(t,v,x) \int_0^1 Y(x,z) \, \frac{dz}{z}$$

$$+ \frac{1}{v} \, [b(t,v,x) - \Phi(t,x)] \; . \qquad (3.57)$$

Note that the intensity in the horizontal direction $b(t,0,x)$ is equal to the function $\Phi(t,x)$,

$$b(t,0,x) = \Phi(t,x) . \tag{3.58}$$

The differential-integral equations (3.53) and (3.57), joined to the differential-integral equations for X, Y, and J,

$$X_x(x,z) = \frac{\lambda}{2} Y(x,z) \int_0^1 Y(z',x) \frac{dz'}{z'} , \tag{3.59}$$

$$Y_x(x,z) = -\frac{1}{z} Y(x,z) + \frac{\lambda}{2} X(x,z) \int_0^1 Y(z',x) \frac{dz'}{z'} , \tag{3.60}$$

$$J_x(t,x,z) = -\frac{1}{z} J(t,x,z) + \frac{\lambda}{2} X(x,z) \int_0^1 J(t,x,z') \frac{dz'}{z'}, \tag{3.61}$$

and the relation

$$\Phi(t,x) = 2 \int_0^1 J(t,x,z') \frac{dz'}{z'} \tag{3.62}$$

form the basis of the numerical procedure. The initial conditions are

$$X(0,z) = 1 , \tag{3.63}$$

$$Y(0,z) = 1 , \tag{3.64}$$

$$J(t,t,z) = \frac{\lambda}{4} X(t,z) , \tag{3.65}$$

for $0 < z \leq 1$ and, when $-1 \leq v < 0$,

$$h(t,v,t) = 0 , \qquad (3.66)$$

$$b(t,v,t) = |v|^{-1} , \qquad (3.67)$$

while when $0 < v \leq 1$ ,

$$h(t,v,t) = \frac{1}{v} Y(v,t) , \qquad (3.68)$$

$$b(t,v,t) = \frac{1}{v} [X(v,t) - 1] . \qquad (3.69)$$

These last equations follow from the physical meanings of b, h, X, and Y.

We shall rederive the Cauchy system for b and h, starting with the integral equation for $\Phi$ . The integral equation may also be written, replacing t by x-t and making a suitable change of variable of integration,

$$\Phi(x-t,x) = \frac{\lambda}{2} E_1(t) + \frac{\lambda}{2} \int_0^x E_1(|t-y|) \, \Phi(y-x) \, dy . \qquad (3.70)$$

The purpose of this pair of moves is to remove x from the forcing function and the kernel. Differentiate this equation with respect to x. This yields

$$\frac{d}{dx} \Phi(x-t,x) = \frac{\lambda}{2} E_1(x-t) \, \Phi(0,x)$$

$$+ \frac{\lambda}{2} \int_0^x E_1(|t-y|) \frac{d}{dx} \Phi(y-x) \, dy . \qquad (3.71)$$

The solution of this integral equation is expressed in terms of the solution of the integral equation when $\frac{\lambda}{2} E_1(x-t)$ is the forcing function,

$$\frac{d}{dx} \Phi(x-t,x) = \Phi(0,x) \; \Phi(t,x) \; . \tag{3.72}$$

This is a differential equation for the function $\Phi$.

We seek a differential equation for $h(t,+v,x)$, $v>0$. Differentiate with respect to $x$ on both sides of the formula for $h$ in terms of $\Phi$ :

$$h(t,+v,x) = \frac{1}{v} \left[ e^{-t/v} + \int_0^t e^{-(t-y)/v} \Phi(x-y,x) \; dy. \right] \tag{3.73}$$

This yields

$$h_x(t,+v,x) = \frac{1}{v} \int_0^t e^{-(t-y)/v} \frac{d}{dx} \Phi(x-y,x) \; dy \; . \tag{3.74}$$

Use of Eq. (3.72) leads to

$$h_x(t,+v,x) = \Phi(0,x) \cdot \frac{1}{v} \int_0^t e^{-(t-y)/v} \Phi(y,x) \; dy \; . \tag{3.75}$$

The second factor is recognized to be $b(t,+v,x)$. In a similar fashion, we derive the equation for $h_x(t,-v,x)$. The basic differential equation for the function $h$ is

$$h_x(t,\pm v,x) = \Phi(0,x) \; b(t,\pm v,x) \; . \tag{3.76}$$

The function $b(t,+v,x)$ may be expressed as

$$b(t,+v,x) = \frac{1}{v} \int_{x-t}^{x} e^{-(t-x+y)/v} \Phi(x-y,x) \, dy \, . \quad (3.77)$$

Differentiation produces the formula

$$b_x(t,+v,x) = \frac{1}{v} e^{-t/v} \Phi(0,x) - \frac{1}{v}\Phi(t,x) + \frac{1}{v} b(t,+v,x)$$

$$+ \frac{1}{v} \int_{x-t}^{x} e^{-(t-x+y)/v} \frac{d}{dx} \Phi(x-y,x) \, dy \, .$$
$$(3.78)$$

The integral is

$$\frac{1}{v} \int_{x-t}^{x} e^{-(t-x+y)/v} \frac{d}{dx} \Phi(x-y,x) \, dy$$

$$= \frac{1}{v} \Phi(0,x) \int_{x-t}^{x} e^{-(t-x+y)/v} \Phi(y,x) \, dy$$

$$= \frac{1}{v} \Phi(0,x) \int_{0}^{t} e^{-(t-y)/v} \Phi(x-y,x) \, dy$$

$$= \Phi(0,x) \left[ h(t,+v,x) - \frac{1}{v} e^{-t/v} \right] . \quad (3.79)$$

The differential equation then takes the form

$$b_x(t,+v,x) = \Phi(0,x) \, h(t,+v,x) + \frac{1}{v} b(t,+v,x)$$

$$- \frac{1}{v} \Phi(t,x) \, . \quad (3.80)$$

A similar procedure is used for $b_x(t,-v,x)$.

The equation for $b(t,\pm v,x)$ is

$$b_x(t,\pm v,x) = \Phi(0,x) \, h(t,\pm v,x)$$

$$\pm \frac{1}{v} \, [b(t,\pm v,x) - \Phi(t,x)] \; . \qquad (3.81)$$

One of the initial conditions requires the evaluation of the expression

$$b(x,+v,x) = \frac{1}{v} \int_0^x e^{-(x-t)/v} \, \Phi(y,x) \, dy \; . \qquad (3.82)$$

Recall the integral equation for the source function $J$,

$$J(t,x,v) = \frac{\lambda}{4} \, e^{-(x-t)/v} + \frac{\lambda}{2} \int_0^x E_1(|t-y|) \, J(y,x,v) \, dy \; . \qquad (3.83)$$

At $t = x$,

$$J(x,x,v) = \frac{\lambda}{4} + \frac{\lambda}{2} \int_0^x E_1(x-y) \, J(y,x,v) \, dy \; . \qquad (3.84)$$

The Hopf lemma enables us to write

$$J(x,x,v) = \frac{\lambda}{4} + \int_0^x \Phi(y,x) \, \frac{\lambda}{4} \, e^{-(x-y)/v} \, dy \; . \qquad (3.85)$$

But

$$J(x,x,v) = \frac{\lambda}{4} \, X(x,v) \; . \qquad (3.86)$$

Then

$$X(x,v) = 1 + \int_0^x \Phi(y,x) \; e^{-(x-y)/v} \; dy \; , \tag{3.87}$$

or

$$\frac{1}{v} [X(x,v) - 1] = \frac{1}{v} \int_0^x \Phi(y,x) \; e^{-(x-y)/v} \; dy \; . \tag{3.88}$$

Therefore, an initial condition at $x = t$ is

$$b(t,+v,t) = \frac{1}{v} [X(t,v) - 1] \; . \tag{3.89}$$

For negative values of the second argument, the initial condition is simply

$$b(t,-v,t) = \frac{1}{v} \; . \tag{3.90}$$

Evaluation of $h(t,+v,x)$ when $t = x$ produces the equation

$$h(x,+v,x) = \frac{1}{v} \int_0^x e^{-(x-y)/v} \Phi(x-y,x) \; dy + \frac{1}{v} e^{-x/v} \; . \tag{3.91}$$

An equation for $J(0,x,v)$ is

$$J(0,x,v) = \frac{\lambda}{4} e^{-x/v} + \frac{\lambda}{2} \int_0^x E_1(y) \; J(y,x,v) \; dy \; , \tag{3.92}$$

and the Hopf lemma produces

$$J(0,x,v) = \frac{\lambda}{4} e^{-x/v} + \int_0^x \Phi(x-y) \frac{\lambda}{4} e^{-(x-y)/v} \, dy \; .$$

$$(3.93)$$

Introduce the function Y by the relation

$$J(0,x,v) = \frac{\lambda}{4} Y(x,v) \; .$$                                $$(3.94)$$

Then we have

$$Y(x,v) = e^{-x/v} + \int_0^x \Phi(x-y) e^{-(x-y)/v} \, dy \qquad (3.95)$$

and

$$h(x,+v,x) = \frac{1}{v} Y(x,v) \; .$$

The initial condition at x = t is

$$h(t,+v,t) = \frac{1}{v} Y(t,v) \; .$$                                $$(3.96)$$

Finally, the condition on $h(t,-v,x)$ at x = t is, from the definition of h,

$$h(t,-v,t) = 0 \; .$$                                $$(3.97)$$


3.4     NUMERICAL METHOD AND RESULTS

        The  b  and  h  functions are readily computed.
First note however, that an examination of the differential
equation for  b  shows that it will be unstable for positive
values of  v  because of the term  $v^{-1} b(t,v,x)$.  This has

been computationally verified. It is for this reason that
the calculation is restricted to negative values of the
direction cosine. The basic differential-integral equations
are Eqs. (3.53),(3.57), (3.59) - (3.61) and (3.62), and the
initial values are given by Eqs. (3.63) - (3.67), in the
preceding section. To obtain b and h for positive
values of v, the equations $b(t,v,x) = h(x-t,-v,x)$ and
$h(t,v,x) = b(x-t,-v,x)$ are used.

The integrals in the set of differential-integral
equations are approximated by sums according to the Gaussian
quadrature formula of order N. Let $w_1, w_2, \ldots, w_N$ be
Christoffel weights and $z_1, z_2, \ldots, z_N$ be the abscissas,
the roots of the shifted Legendre polynomial of degree N.
For N = 7, these are given in Appendix A.

Let us suppose that it is desired to compute
$b(t,-v,x)$ and $h(t,-v,x)$ for a specified value of $-v$
where $-1 \leq -v < 0$, for thicknesses x = a, 2a, 3a, ..., Ka,
where Ka is the maximum desired thickness. For each thickness the altitudes are to be t = 0, a, 2a, ..., x. For
b and h at the bottom, the formulas are

$$b(0,-v,x) = h(x,v,x) = \frac{1}{v} Y(x,v) , \qquad (3.98)$$

$$h(0,-v,x) = b(x,v,x) = \frac{1}{v} [X(x,v)-1] , \qquad (3.99)$$

and, at the top,

$$b(x,-v,x) = \frac{1}{v} , \qquad (3.100)$$

$$h(x,-v,x) = 0 . \qquad (3.101)$$

For  b  and  h  with positive direction cosines, recall that
the formulas are

$$b(t,v,x) = h(x-t,-v,x) \; , \tag{3.102}$$

$$h(t,v,x) = b(x-t,-v,x) \; . \tag{3.103}$$

Let  X, Y,  and  J  be evaluated for the direction
cosines  $z_1, z_2, \ldots, z_N$  and put  $X_i(x) = X(x,z_i)$, $Y_i(x)$
$= Y(x,z_i)$, $J_i(t,x) = J(t,z_i,x)$.  Since  v  is arbitrary, let
us choose  $v = z_1, z_2, \ldots, z_N$.  This makes it so that the
needed values of  X  and  Y  in Eqs. (3.98) and (3.99) are
just the ones being computed.  Thus we write  $b_i(t,x)$
$= b(t,-z_i,x)$, $h_i(t,x) = h(t,-z_i,x)$.  Finally put
$t = t_j = ja$  for  $j = 1, 2, \ldots,$  and write

$$J_{ij}(x) = J_i(t_j,x) = J(t_j,z_i,x) \; , \tag{3.104}$$

$$b_{ij}(x) = b_i(t_j,x) = b(t_j,-z_i,x) \; , \tag{3.105}$$

$$h_{ij}(x) = h_i(t_j,x) = h(t_j,-z_i,x) \; . \tag{3.106}$$

The basic equations for the computational procedure are

$$X_i' = \frac{\lambda}{2} Y_i \sum_{k=1}^{N} Y_k \frac{w_k}{z_k} \; , \tag{3.107}$$

$$X_i(0) = 1 \; , \tag{3.108}$$

$$Y_i' = -\frac{1}{z_i} Y_i + \frac{\lambda}{2} X_i \sum_{k=1}^{N} Y_k \frac{w_k}{z_k} \; , \tag{3.109}$$

$$Y_i(0) = 1 , \tag{3.110}$$

$$J'_{ij} = \frac{-1}{z_i} J_{ij} + \frac{\lambda}{4} X_i \Phi_j , \tag{3.111}$$

$$J_{ij}(t_j) = \frac{\lambda}{4} X_i(t_j) , \tag{3.112}$$

$$h'_{ij} = \frac{\lambda}{2} b_{ij} \sum_{k=1}^{N} Y_k \frac{w_k}{z_k} , \tag{3.113}$$

$$h_{ij}(t_j) = 0 , \tag{3.114}$$

$$b'_{ij} = \frac{\lambda}{2} h_{ij} \sum_{k=1}^{N} Y_k \frac{w_k}{z_k} - z_i^{-1} [b_{ij} - \Phi_j] , \tag{3.115}$$

$$b_{ij}(t_j) = \frac{1}{z_i} , \tag{3.116}$$

$$i = 1, 2, \ldots, N; \quad j = 1, 2, \ldots, N,$$

where

$$\Phi_j = 2 \sum_{k=1}^{N} J_{kj} \frac{w_k}{z_k} , \quad j = 1, 2, \ldots . \tag{3.117}$$

The computation begins with the 2N differential equations for X and Y, Eqs. (3.107) and (3.109) and the initial values at $x = 0$ in Eqs. (3.108) and (3.110). These 2N equations are integrated to $x = a = t_1$. At this point, values of $X(z_i,a)$ and $Y(z_i,a)$ are known. The quantities $J(0,z_i,a)$, $J(a,z_i,a)$, $b(0,-z_i,a)$, $b(a,-z_i,a)$, $h(0,-z_i,a)$, $h(a,-z_i,a)$, $b(0,+z_i,a)$, $h(z,+z_i,a)$, $h(0,+z_i,a)$, $h(a,+z_i,a)$, and $\Phi(0,a)$ may all be evaluated using the relevant formulas. The desired quantities b, h, $\Phi$, ... are printed out.

The calculation continues.  The 3N  initial conditions of  Eqs. (3.112), (3.114), and (3.116) are imposed
for  j = 1  and  i = 1, 2, ..., N.  The  3N  differential
equations  (3.111), (3.113), and (3.115) are adjoined to the
existing set.  The enlarged system is integrated until
$x = 2a = t_2$.  Some of the desired quantities are known at
this point, others are computed via simple formulas.  Print-
out of the required quantities is done.  The procedure continues in this way until  x = Ka.  The number of differential equations being integrated from  x = (K-1)a  to
x = KA  is  2N + (K-1)3N.  For  N = 7  and  K = 50,  this
number is only 1043.

A program written in FORTRAN IV is listed in Appendix B.  It was used to produce tables of  b, h and $\Phi$ ,
given in Appendix C.  For those tables, quadrature formulas
of order  N = 7  are used, together with an integration
step size of 0.005 and an Adams-Moulton integration method.
We checked for consistency of results using an increased
value of  N, N = 9,  and a simultaneously decreased step
length of 0.0025.  Generally, five figures of agreement were
found.  The discrepancy was always less than about 0.02
percent.  Other independent calculations agree with these
results.

Some graphical results for  b  are shown in Figs.
3.4 through 3.10.  Figures 3.4 through 3. 7 show internal
intensity patterns at various altitudes in slabs having the
following albedos and thicknesses; $\lambda = 0.1$, x = 2; $\lambda = 0.5$,
x = 2; $\lambda = 1.0$, x = 5; $\lambda = 1.0$, x = 20.  The horizontal
axis is the polar angle.  "Up" refers to upwelling directions
and the polar angle is measured from the upward-directed

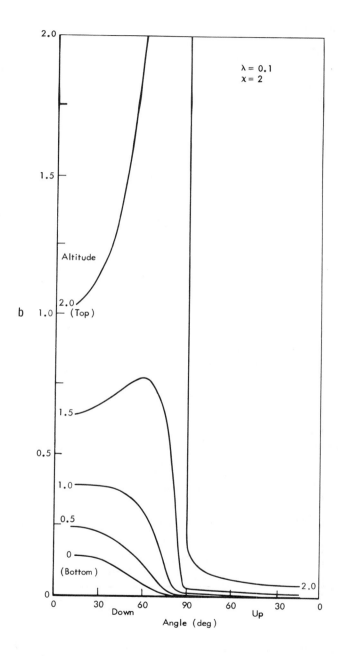

Figure 3.4   Intensity Patterns   b   for   $\lambda = 0.1$, $x = 2$

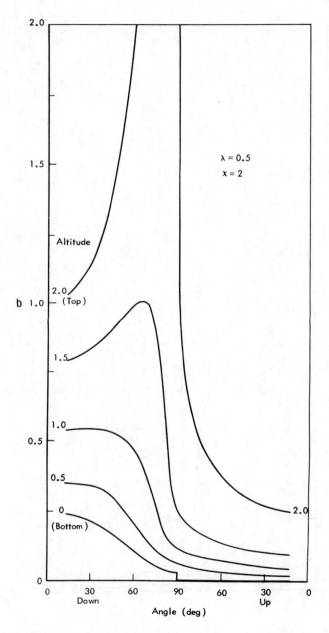

Figure 3.5   Intensity Patterns   b   for   $\lambda = 0.5$, $x = 2$

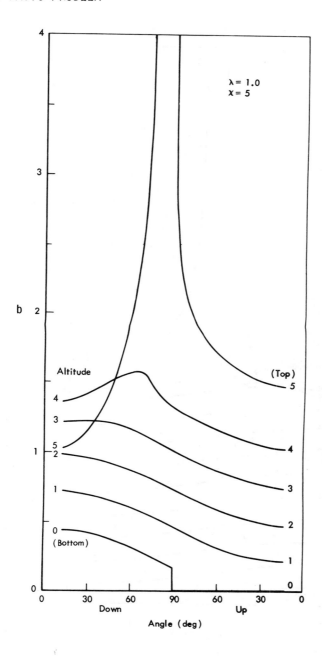

Figure 3.6   Intensity Patterns   b   for   λ = 1.0, x = 5

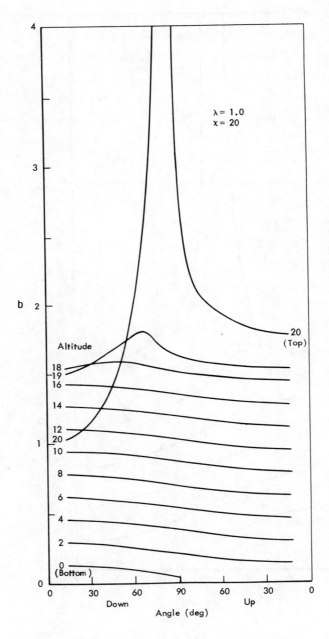

Figure 3.7  Intensity Patterns  b   for  $\lambda = 1.0$, x = 20

vertical. "Down" refers to downwelling directions and the polar angle is measured from the downward-directed vertical.

Figures 3.8 through 3.10 show the depth variation of the intensities in fixed directions. Figure 3.8 is a plot for the case $\lambda = 0.1$, $x = 2$ and the three directions are $\sim 13°$ in the upper hemisphere, the horizontal direction, and $\sim 13°$ in the downwelling direction. Each curve on the logarithmic plot is nearly a straight line over most depths, indicating an exponential dependence on depth. Figure 3.9 is a composite graph, showing b for the albedo 0.7, thicknesses 2.0 and 5.0. Corresponding curves for the two thicknesses overlap to some degree. Of course the curves for thickness 2.0 extend only to depth 2.0. Figure 3.10 is a composite graph of b for the conservative case and thicknesses 5, 10 and 20. The dependence of b on depth is shown to be linear for depths greater than approximately 2.

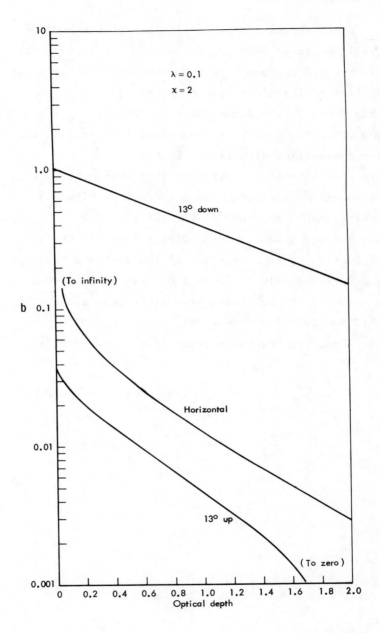

Figure 3.8   Intensities   b   as Functions of Depth for
              $\lambda = 0.1$, $x = 2$

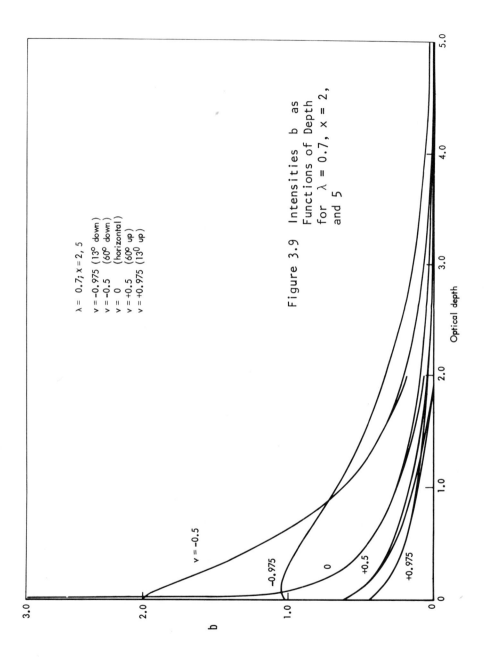

λ = 0.7; x = 2, 5

v = −0.975 (13° down)
v = −0.5    (60° down)
v = 0        (horizontal)
v = +0.5    (60° up)
v = +0.975 (13° up)

Figure 3.9   Intensities  b  as
Functions of Depth
for  λ = 0.7,  x = 2,
and 5

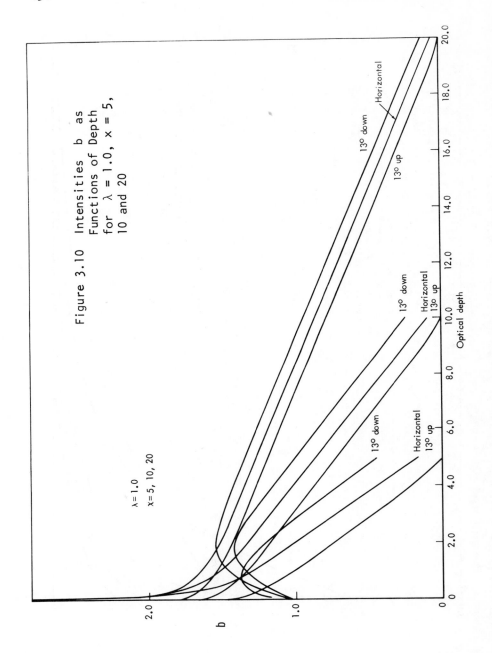

Figure 3.10   Intensities b as
Functions of Depth
for λ = 1.0, x = 5,
10 and 20

## EXERCISES

1.  Prove that the functions b and h which are the
    solutions of the initial value problem are also the
    functions which are defined in Eqs. (3.7) and (3.14).

2.  Derive a system of integral equations which are
    satisfied by the functions b and h.

3.  The FORTRAN program for computing b and h is
    listed in Appendix B. Compute tables of the func-
    tions b and h like those in Appendix C.

4.  Use tables of b and h functions to calculate X,
    Y, J, and I.

## REFERENCES

1.  H. Kagiwada and R. Kalaba, "A New Initial Value Method
    for Internal Intensities in Radiative Transfer,"
    *Astrophysical Journal*, v. 147, 1967, pp. 301-309.

2.  H. Kagiwada and R. Kalaba, "The Equivalence of the
    Isotropic and Monodirectional Source Problems," *J.
    Quant. Spectrosc. Rad. Transfer*, v. 8, 1968, pp. 843-846.

3.  H. Kagiwada and R. Kalaba, *Numerical Results for Internal
    Intensities in Atmospheres Illuminated by Isotropic
    Sources*, The Rand Corporation, RM-4958-PR, 1966.

4.  H. Kagiwada and R. Kalaba, *Integral Equations via Imbed-
    ding Methods*, Addison-Wesley Publishing Co., Reading,
    Mass., 1974.

CHAPTER 4

INTERNAL SOURCES

## 4.1   INTRODUCTION

Multiple scattering in an atmosphere containing
emitting sources of radiation is one of the central topics
in meteorology.  It also appears in the astrophysical problems
of stellar spectra and diffusion of radiation in the galaxy,
as well as in neutron transport and heat transfer.  The equations which are developed here lead to an effective computational procedure for the determination of internal intensities
and source functions, regardless of the nature of the geo-
metric variation of emission rate with height in a homogeneous
slab atmosphere.  Similar results may be obtained for inhomogeneous and anisotropic atmospheres, and for other
generalizations.

We consider anew the problem of determining the radiation field produced within a finite homogeneous slab which
both absorbs radiation and scatters it isotropically, with
isotropic sources of radiation being located in the slab and
their strength depending only on the altitude above the

bottom. In this chapter, we derive three different initial value problems for the computation of the internal and external radiation fields. The first produces internal intensity and source functions directly. The second yields the resolvent as the solution of a Cauchy system. The third method is to produce a part of the resolvent, namely the function $\Phi(t,x)$, and expresses the solution by the use of a representation formula.

While the statements in the preceding paragraph are made for direct problems, we also consider inverse problems in this chapter. By an inverse problem we mean the estimation of properties of a scattering and absorbing medium and of the nature of the sources, based on measurements of a part of the radiation field. In particular we discuss the estimation of the source distribution in an emitting atmosphere using emergent intensity measurements made for different emergent angles. We pose this inverse problem as a nonlinear boundary value problem. The technique called quasilinearization is effective in the treatment of such problems. This is described, then applied to the problem at hand, and we present some numerical results which display the effectiveness of this procedure. Other types of inverse problems are discussed in succeeding chapters.

## 4.2    CAUCHY SYSTEM FOR SOURCE FUNCTIONS, AND EMERGENT AND INTERNAL INTENSITIES

Consider a homogeneous slab of optical thickness x which absorbs radiation and reradiates it isotropically. The albedo for single scattering is denoted by the constant

$\lambda$ .   Located within the slab are isotropic sources of radiation.   The emission of energy per unit volume per unit solid
angle per unit time at altitude   t   is   $g(t)$ ,   $0 \leq t \leq x$ .   A
radiation field is created, and we wish to determine intensities and rates of production of energy in the slab.

The source function   $u(t,x)$   is the total rate of
production of energy per unit volume per unit solid angle
at altitude   t   in the described slab of thickness   x .   For
brevity, we may refer to   $u(t,x)$   as the rate of production
of energy, assuming a unit volume and unit solid angle.   The
source function satisfies the integral equation

$$u(t,x) = g(t) + \frac{\lambda}{2} \int_0^x E_1(|t-y|) \; u(y,x) \; dy ,$$

$$0 \leq t \leq x \leq x_1 . \quad (4.1)$$

The intensity of the radiation emerging with polar angle
whose cosine is   v   is

$$e(v,x) = v^{-1} \int_0^x e^{-(x-y)v} u(y,x) dy ,$$

$$0 \leq v \leq 1 . \quad (4.2)$$

Let us derive, physically, a system of differential
equations and initial conditions for the determination of the
source function and some auxiliary functions.   We consider the
rate of production of energy per unit solid angle per unit
volume at altitude   t   in a slab of thickness   x .   We increase the thickness of the slab by the addition of a layer
of thickness   $\Delta$.   The rate of production of energy at
altitude   t   in the enlarged slab is

$$u(t,x+\Delta) = u(t,x) + u(x,x) \; \Delta\Phi(t,x) + o(\Delta) . \quad (4.3)$$

The second term is due to isotropic scattering at the top
followed by the production of energy at altitude  t.   The
rate of production of energy in a volume of unit base area
and height  $\Delta$  at the top is expressible as

$$u(x,x) \ \Delta = \left[g(x) + \frac{\lambda}{2}\int_0^1 e(v,x) \ dv\right]\Delta \ , \qquad (4.4)$$

and so

$$u(x,x) = \left[g(x) + \frac{\lambda}{2}\int_0^1 e(v,x) \ dv\right]. \qquad (4.5)$$

The function  $\Phi(t,x)$  represents the rate of production of
energy at altitude  t  in a slab of thickness  x  due to
isotropic sources at the top, and it may be written,

$$\Phi(t,x) = \frac{\lambda}{2}\int_0^1 J(t,x,v) \ dv/v \ . \qquad (4.6)$$

By the addition of a slab of thickness  $\Delta$  to the top
of a slab of thickness  x ,  we have the relation,

$$e(v,x+\Delta) = e(v,x) - e(v,x)\Delta \ v^{-1} + v^{-1} \ X(x,v) \ u(x,x)\Delta$$
$$+ \ o(\Delta) \ . \qquad (4.7)$$

By passage to the limit  $\Delta \to 0$  in Eqs. (4.3) and (4.7), we
obtain the differential equations

$$u_x(t,x) = \left[g(x) + \frac{\lambda}{2}\int_0^1 e(v',x) \ dv'\right]$$
$$- \frac{\lambda}{2}\int_0^1 J(t,x,v'') \ dv''/v'' \ , \qquad 0 \leq t \leq x \ , \qquad (4.8)$$

$$e_x(v,x) = -v^{-1} e(v,x)$$

$$+ v^{-1} X(x,v) \left[ g(x) + \frac{\lambda}{2} \int_0^1 e(v',x) \, dv' \right] ,$$

$$0 \leq v \leq 1 , \quad (4.9)$$

for $0 \leq x \leq x_1$ .

We consider the slab of thickness $\Delta$ in the limit as $\Delta$ tends to 0, and we write the initial condition,

$$e(v,0) = 0 . \qquad (4.10)$$

As for $u(t,x)$ evaluated at $x = t$, it must simply be $u(x,x)$ evaluated at $x = t$, or

$$u(t,t) = g(t) + \frac{\lambda}{2} \int_0^1 e(v',x) \, dv' , \qquad (4.11)$$

and this is the initial condition on $u(t,x)$. This system of differential equations (4.8) and (4.9) and initial conditions (4.10) and (4.11) must be augmented by equations for $X$, and $Y$, and $J$ :

$$X_x(x,z) = \frac{\lambda}{2} Y(x,z) \int_0^1 Y(x,z') \, dz'/z' , \qquad (4.12)$$

$$Y_x(x,z) = -z^{-1} Y(x,z) + \frac{\lambda}{2} X(x,z) \int_0^1 Y(x,z') dz'/z' , \qquad (4.13)$$

$$J_x(t,x,z) = -z^{-1} J(t,x,z)$$

$$+ \lambda X(x,z) \int_0^1 J(t,x,z') \, dz'/z' , \qquad (4.14)$$

$$X(0,z) = 1 \; , \tag{4.15}$$

$$Y(0,z) = 1 \; , \tag{4.16}$$

$$J(t,t,z) = \frac{\lambda}{4} X(t,z) \; . \tag{4.17}$$

Let us rederive the initial value problem for $u(t,x)$ and $e(v,x)$ analytically from the integral equation for $u$. Differentiation on both sides of Eq. (4.1) yields

$$u_x(t,x) = \frac{\lambda}{2} E_1(x-t) \; u(x,x)$$

$$+ \frac{\lambda}{2} \int_0^x E_1(|t-y|) \; u_x(y,x) \; dy \; . \tag{4.18}$$

Recognizing that $\Phi(t,x)$ is the solution of the integral equation

$$\Phi(t,x) = \frac{\lambda}{2} E_1(x-t) + \frac{\lambda}{2} \int_0^x E_1(|t-y|) \; \Phi(y,x) \; dy \; , \tag{4.19}$$

we express the solution of the integral Eq. (4.18), using the superposition principle as

$$u_x(t,x) = u(x,x) \; \Phi(t,x) \; . \tag{4.20}$$

We already know the Cauchy problem for $\Phi(t,x)$. Let us now consider $u(x,x)$ ,

$$u(x,x) = g(x) + \frac{\lambda}{2} \int_0^x E_1(x-y) \, u(y,x) \, dy$$

$$= g(x) + \frac{\lambda}{2} \int_0^x \left( \int_0^1 e^{-(x-y)/v} dv/v \right) u(y,x) \, dy$$

$$= g(x) + \frac{\lambda}{2} \int_0^1 \left( v^{-1} \int_0^x e^{-(x-y)/v} u(y,x) \, dy \right) dv \; .$$

$$(4.21)$$

(We assume that the interchange in the order of integration is valid.) Defining the function $e(v,x)$ ,

$$e(v,x) = v^{-1} \int_0^x e^{-(x-y)/v} \, u(y,x) \, dy \; , \qquad (4.22)$$

then we express $u(x,x)$ as

$$u(x,x) = g(x) + \frac{\lambda}{2} \int_0^1 e(v,x) \, dv \; . \qquad (4.23)$$

To derive a differential equation for $e(v,x)$ we differentiate on both sides of Eq. (4.22),

$$e_x(u,x) = v^{-1} \left\{ u(x,x) - v^{-1} \int_0^x e^{-(x-y)/v} u(y,x) \, dy \right.$$

$$\left. + \int_0^x e^{-(x-y)/v} u_x(y,x) \, dy \right\}$$

$$= v^{-1} \left\{ u(x,x) - e(v,x) \right.$$

$$\left. + \int_0^x e^{-(x-y)/v} u(x,x) \, \Phi(y,x) \, dy \right\}$$

$$= -v^{-1} e(v,x)$$

$$+ v^{-1} u(x,x) \left\{ 1 + \int_0^x e^{-(x-y)/v} \Phi(y,x) \, dy \right\}$$

$$= - v^{-1} e(v,x) + v^{-1} u(x,x) \, X(x,v) \, . \qquad (4.24)$$

The initial conditions result from setting $x = 0$ in the definition of $e(v,x)$, and evaluating $u(x,t)$ at $x = t$ ,

$$e(v,0) = 0 \, , \qquad\qquad\qquad\qquad\qquad\qquad (4.25)$$

$$u(t,t) = g(t) + \frac{\lambda}{2} \int_0^1 e(v,t) \, dv \, . \qquad\qquad (4.26)$$

Next, we wish to determine the total intensity of radiation at the fixed altitude $t$, $0 \le t \le x$ , which is propagating in a direction whose direction cosine with respect to the upward vertical is $v$. This intensity is denoted by $I = I(t,v;x)$, which draws attention to the fact that $I$ is to be considered a function of $t$, $v$, and $x$ , the optical thickness. As usual, the intensity at a

particular point for a particular direction is the energy
per unit time per unit normal area per unit solid angle.

Let  $I(t,v,x)$  be the total intensity at altitude
t  in a direction whose direction cosine is  v  measured
from the upward vertical  $(-1 \leq v \leq +1)$,  the slab thickness
being  x,  and there being sources with distribution  $g(y)$,
$(0 \leq t, y \leq x)$.

A differential equation for  I  is derived in the
usual manner. Add a slab of thickness  $\Delta$  to the top of
the slab of thickness  x  and relate the intensities in the
two slabs at altitude  t  and with direction cosine  v.
The equation is

$$I(t,v,x+\Delta) = I(t,v,x) + u(x,x)\Delta b(t,v,x) + o(\Delta) ,$$

$$(4.27)$$

where  $u(x,x)$  is the source function at the top, given by
Eq. (4.5), and  $b(t,v,x)$  is the total intensity at altitude
t  in a direction whose direction cosine is  v  measured
from the upward vertical  $(-1 \leq v \leq +1)$ ,  the slab thickness
being  x,  and there being omnidirectional sources at the top.

More precisely, the omnidirectional sources are such
that unit energy per unit horizontal area per unit time is
incident at the top surface of the slab in all downward
directions. Recall that total intensity refers to all of
the energy at a given place in a given direction, whether it
has interacted with the medium one or more times, or not at
all. In contrast, diffuse intensity refers to the energy
which has been emitted in the medium, or scattered one or
more times.

In the limit as  $\Delta$  tends to 0, there results the differential equation,

$$I_x(t,v,x) = b(t,v,x) \left[ g(x) + \frac{\lambda}{2} \int_0^1 e(v',x) \, dv' \right] \tag{4.28}$$

for $x \geq t$. For an initial condition at $x = t$, there is

$$I(t,v,t) = e(v,t) , \qquad\qquad 0<v\leq 1 , \qquad (4.29)$$

or

$$I(t,v,t) = 0 , \qquad\qquad -1\leq v<0 . \qquad (4.30)$$

The complete system of differential equations for determining $I(t,v,x)$ is

$$e_x(v,x) = -v^{-1} e(v,x)$$

$$+ v^{-1}X(v,x)\left[g(x) + \frac{\lambda}{2}\int_0^1 e(v',x) \, dv'\right] , \tag{4.31}$$

$$X_x(x,v) = \frac{\lambda}{2} Y(x,v) \int_0^1 Y(x,v') \, dv'/v' , \tag{4.32}$$

$$Y_x(x,v) = -v^{-1}Y(x,v) + \frac{\lambda}{2} X(x,v) \int_0^1 Y(x,v') \, dv'/v' , \tag{4.33}$$

$$b_x(t,v,x) = \frac{\lambda}{2} h(t,v,x) \int_0^1 Y(x,v') \, dv'/v'$$

$$+ v^{-1}[b(t,v,x) - \Phi(t,x)], \tag{4.34}$$

$$h_x(t,v,x) = \frac{\lambda}{2} b(t,v,x) \int_0^1 Y(x,v') \, dv'/v' \, , \qquad (4.35)$$

$$J_x(t,x,z) = - z^{-1} J(t,x,z) + \frac{\lambda}{2} X(x,z) \, \Phi(t,x) \, , \qquad (4.36)$$

$$I_x(t,v,x) = b(t,v,x) \left[ g(x) + \frac{\lambda}{2} \int_0^1 e(v',x) \, dv' \right] , \qquad (4.37)$$

where

$$\Phi(t,x) = 2 \int_0^1 J(t,x,z') \, dz'/z' \, , \qquad (4.38)$$

with the initial conditions,

$$e(v,0) = 0 \, , \qquad\qquad\qquad 0<v\leq1 \, , \qquad (4.39)$$

$$X(0,u) = 1 \, , \qquad\qquad\qquad 0<v\leq1 \, , \qquad (4.40)$$

$$Y(0,u) = 1 \, , \qquad\qquad\qquad 0<v\leq1 \, , \qquad (4.41)$$

$$b(t,v,t) = - v^{-1} \, , \qquad\qquad\qquad -1<v<0 \, , \qquad (4.42)$$

$$h(t,v,t) = 0 \, , \qquad\qquad\qquad -1<v<0 \, , \qquad (4.43)$$

$$J(t,t,z) = \frac{\lambda}{4} X(t,z) \, , \qquad\qquad\qquad 0<z\leq1 \, , \qquad (4.44)$$

$$I(t,v,t) = e(v,t) \, , \qquad\qquad\qquad 0<v\leq1 \, , \qquad (4.45)$$

$$I(t,v,t) = 0 \, , \qquad\qquad\qquad -1\leq v<0 \, . \qquad (4.46)$$

The functions  b  and  h  satisfy the relations

$$b(t,v,x) = h(x-t, -v, x) \ , \tag{4.47}$$

$$h(t,v,x) = b(x-t, -v, x) \ . \tag{4.48}$$

## 4.3    NUMERICAL METHOD AND EXAMPLES

Let us describe the numerical procedures for calculating  $u(t,x)$  based on Eqs. (4.8) through (4.17). Integrals are to be approximated by sums according to a quadrature formula of order  N.  Let

$$u(x) = u(t,x) \ ,$$

$$e_i(x) = e(v_i,x) \ ,$$

$$J_i(x) = J(t,x,v_i) \ ,$$

$$X_i(x) = X(x,v_i) \ ,$$

$$Y_i(x) = Y(x,v_i) \ , \qquad\qquad i = 1,2,\ldots,N \ , \tag{4.49}$$

where  $\{v_i\}$  are the  N  quadrature points, and  $\{w_i\}$  will denote the weights.  The approximate system of differential equations, is

$$X_i' = \frac{\lambda}{2} Y_i \sum_{k=1}^{N} Y_k \, w_k/v_k \ , \tag{4.50}$$

$$Y_i' = - v_i^{-1} Y_i + \frac{\lambda}{2} X_i \sum_{k=1}^{N} Y_k \, w_k/v_k \ , \tag{4.51}$$

$$e_i' = - v_i^{-1} e_i$$

$$+ v_i^{-1} X_i \left[ g(x) + \frac{\lambda}{2} \sum_{k=1}^{N} e_k w_k \right] , \tag{4.52}$$

$$J_i' = - v_i^{-1} J_i + \lambda X_i \sum_{k=1}^{N} J_k w_k / v_k , \tag{4.53}$$

$$u' = \left[ g(x) + \frac{\lambda}{2} \sum_{k=1}^{N} e_k w_k \right] \frac{\lambda}{2} \sum_{k=1}^{N} J_k w_k / v_k , \tag{4.54}$$

and the initial conditions are

$$X_i(0) = 1 , \tag{4.55}$$

$$Y_i(0) = 1 , \tag{4.56}$$

$$e_i(0) = 0 , \tag{4.57}$$

$$J_i(t) = \frac{\lambda}{4} X_i(t) , \tag{4.58}$$

$$u(t) = g(t) + \frac{\lambda}{2} \sum_{k=1}^{N} e_k w_k . \tag{4.59}$$

At the start of the computation, we set up the initial conditions of Eqs. (4.55) through (4.57). We numerically integrate the 3N differential Eqs. (4.50) - (4.52) for $0 < x \leq t$ . At the point $x = t$, we impose the values in Eqs. (4.58) and (4.59). We adjoin the N+1 differential Eqs. (4.53) and (4.54) and integrate the 4N+1 equations for $t < x \leq x_1$ , where $x_1$ is the maximum value of thickness. In this way we produce values of these functions for a whole

range of thicknesses.   The equations are stable and the
results are accurately and quickly produced.

    We present some numerical results for thickness
x = 1.5 and albedo  $\lambda$ = 1.0.   For the case of a constant
distribution of sources,  g(t) $\equiv$ 1.0, the emergent intensity
pattern appears as in Fig. 4.1.

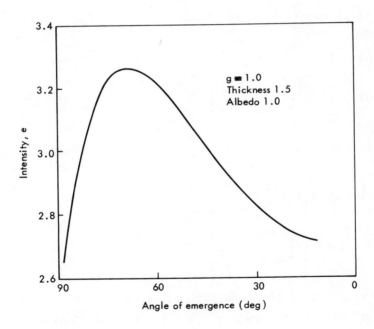

Fig. 4.1   Emergent Intensity for a Slab as a Function of
           Angle of Emergence with   g $\equiv$ 1

Next, with the source distribution function

$$g(t) = 0.5 \ [\tanh 10 \ (t-0.4) - \tanh 10 \ (t-0.7)] \ ,$$

$$(4.60)$$

as plotted in Fig. 4.2, the emergent intensity is that shown
in Figure 4.3.  These solutions were determined by numerical
integration of the approximate system of ordinary differential equations which result by the use of Gaussian quadrature
with seven points.

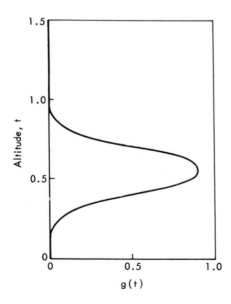

Figure 4.2  A Source Distribution Function

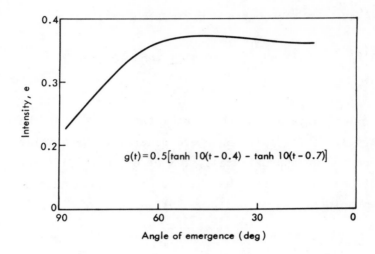

Figure 4.3   Emergent Intensity for a Slab as a Function of
              Angle of Emergence


## 4.4      INVERSE PROBLEMS FOR ESTIMATION OF SOURCE DISTRIBUTION

In this section, we consider inverse problems in which
the task is to estimate the distribution of sources in a plane
parallel atmosphere based on observations of emergent intensities.   Such inverse problems are conceptually similar to
system identification problems of control theory.   A variety
of techniques are now available.   For the most part, we shall
deal with the quasilinearization method which has been found
to be useful in many applications.

The emergent intensities and  X  and  Y  functions
evaluated at the ordinates for Gaussian quadrature with 7

points, i.e., the functions

$$e_i(x) = e(v_i, x) ,$$

$$X_i(x) = X(x, v_i) ,$$

$$Y_i(x) = Y(x, v_i) , \qquad i = 1, 2, \ldots, 7 , \qquad (4.61)$$

satisfy the system of ordinary differential equations,

$$e_i'(x) = v_i^{-1} \left\{ - e_i + X_i g(x) + \frac{\lambda}{2} X_i \sum_{j=1}^{7} e_j w_j \right\} , \qquad (4.62)$$

$$X_i'(x) = \frac{\lambda}{2} Y_i \sum_{j=1}^{7} Y_j w_j / v_j , \qquad (4.63)$$

$$Y_i'(x) = - v_i^{-1} Y_i + \frac{\lambda}{2} X_i \sum_{j=1}^{7} Y_j w_j / v_j . \qquad (4.64)$$

and initial conditions,

$$e_i(0) = 0 , \qquad (4.65)$$

$$X_i(0) = 1 , \qquad (4.66)$$

$$Y_i(0) = 1 , \qquad i = 1, 2, \ldots, 7 . \qquad (4.67)$$

Suppose that we know the thickness to be $x_1 = 1.5$ , and that we believe that the source distribution has the form

$$g(t) = 1.5 [\tanh 10(y-a) - \tanh 10(y-b)] ,$$
$$0 \leq t \leq 1.5 , \qquad (4.68)$$

where the constants  a  and  b  are as yet unknown.  We wish
to determine both  a  and  b  by fitting the observations
which are approximately

$$e_i(1.5) \approx \beta_i , \qquad (i = 1, 2, \ldots, 7) \qquad (4.69)$$

in a least squares sense, that is, we wish to obtain

$$\min S , \qquad (4.70)$$

where

$$S = \sum_{i=1}^{7} [e_i(1.5) - \beta_i]^2 . \qquad (4.71)$$

This inverse problem is a nonlinear boundary value
problem - nonlinear because the unknowns, the $\{e_i\}$ , $\{x_i\}$ ,
$\{Y_i\}$  and  a  and  b,  appear nonlinearly in the differential equations, and a boundary value problem because conditions are given at both ends,  x = 0  and  x = 1.5.  This
is a problem which may be solved by quasilinearization.


## 4.5    QUASILINEARIZATION

Quasilinearization is a rapidly convergent successive
approximation method suitable for the solution of nonlinear
boundary value problems by a high speed computer.  Consider
the general type of nonlinear boundary value problem for the
N-dimensional vector function  u(x)  whose components are
$u_i(x)$, i = 1, 2, ..., N.  The differential equations,

$$u_i'(x) = f_i(u) , \qquad i = 1, 2, \ldots, N;$$
$$0 \leq x \leq x_1 . \qquad (4.72)$$

are to be solved subject to the boundary conditions,

$$u_i(0) = r_i , \qquad i = 1, 2, \ldots, M , \qquad (4.73)$$

$$u_i(x_1) = s_i , \qquad i = M+1, M+2, \ldots, N , \qquad (4.74)$$

where $M < N$. One way to solve this problem is to guess that the missing initial conditions are $s_i^*$,

$$\tilde{u}_i \sim s_i^* , \qquad i = M+1, M+2, \ldots, N , \qquad (4.75)$$

and numerically integrate the nonlinear equations from $x = 0$ to $x = x_1$. However one cannot be at all sure that the terminal conditions will be satisfied at $x = x_1$. A more systematic procedure is required.

Let us apply Newton's method for linearizing a non-linear function of $u$, $f(u)$, by expanding using a current approximation to $u$, $u^o(x)$. Recall the Taylor series

$$f(u) = g(u^o) + f'(u^o)(u - u^o) + \frac{1}{2} f''(u^o) (u - u^o)^2$$
$$+ \ldots . \qquad (4.76)$$

The next approximation to the solution (denoted $u^1(x)$) is obtained by truncating after the linear term ,

$$f(u^1) \cong f(u^o) + f'(u^o) (u^1 - u^o) . \qquad (4.77)$$

In vector matrix form, the above equation becomes

$$f(u^1) = f(u^0) + J(u^0) \ (u^1 - u^0) \ , \tag{4.78}$$

where the Jacobian matrix $J(u^0)$ has elements

$$J_{ij}(u^0) = (\partial / \partial \ u_j^0) \ f_i(u^0) \ . \tag{4.79}$$

In quasilinearization, we replace the original non-linear differential equations by the approximate linear system

$$u_i^1(x) = f_i(u^0) + \sum_{k=1}^{N} \ [\partial f_i(u^0)/\partial u_k^0] \ (u_k - u_k^0) \ , \tag{4.80}$$

where for ease of notation we delete any superscript on the new approximation of $u(x)$. We now have a linear boundary value problem to solve.

To solve a linear boundary value problem, we recall the superposition principle that expresses the solution in terms of a particular solution and a complete set of independent solutions of the homogeneous system. We write

$$u_i(x) = p_i(x) + \sum_{j=1}^{N-M} \ \alpha_j \ h_{ij}(x) \ , \tag{4.81}$$

where $\left\{p_i(x)\right\}$ is a particular solution, the columns of the matrix $\left\{h_{ij}(x)\right\}$ form $N-M$ independent solutions of the homogeneous system and $\left\{\alpha_j\right\}$ are constants. The particular solution is produced numerically by solving the initial value problem

$$p_i'(x) = f_i(u^0) + \sum_{k=1}^{N} [\partial f_i(u^0)/\partial u_k^0] (p_k - u_k^0) ,$$

$$i = 1, 2, \ldots, N \quad , \quad 0 \leq x \leq x_1; \quad (4.82)$$

$$p_i(0) = \begin{cases} r_i, & i = 1, 2, \ldots, M , \\ \\ 0 , & i = M+1, M+2, \ldots, N . \end{cases} \quad (4.83)$$

The homogeneous solutions satisfy the initial value problems

$$h_{ij}'(x) = \sum_{k=1}^{N} [\partial f_i(u^0)/\partial u_k^0] h_{kj} ,$$

$$i = 1, 2, \ldots, N; \quad j = 1, 2, \ldots, N-M ,$$

$$0 \leq x \leq x_1 , \quad (4.84)$$

$$h_{ij}(0) = \begin{cases} 0 , & (i \neq M+j) , \\ \\ 1, & (i = M+j) , \end{cases}$$

$$j = 1, 2, \ldots, N-M. \quad (4.85)$$

The choice of initial conditions on $\{p_i\}$ and $\{h_{ij}\}$ assures us that the initial conditions on $\{u_i\}$ are met (see Eqs. (4.73)).

The N-M constants $\{\alpha_j\}$ are to be chosen so as to satisfy the terminal conditions, Eqs. (4.74), namely that

$$p_i(x_1) + \sum_{j=1}^{N-M} \alpha_j h_{ij}(x_1) = s_i , \quad i = M+1, M+2, \ldots, N .$$

$$(4.86)$$

In the above equations, the functions $\{p_i(x_1)\}$ and $\{h_{ij}(x_1)\}$ are known numerically. Equations (4.86) form a set of N-M linear algebraic equations for N-M unknowns. By solving this system we determine the multipliers which appear in Eq. (4.81), and we can produce the new approximation to $u(x)$ for $0 \leq x \leq x_1$.

This procedure requires that an initial approximation be produced. At any one stage, only the previous approximation is stored, and the particular and homogeneous solutions are calculated. After the constants have been determined, the required linear combination of complementary solutions is produced and stored in place of the previous approximation. The procedure continues in this way for a fixed number of approximations or until a certain degree of consistency is obtained. Since the method is nearly quadratically convergent, only about four approximations are needed when the initial approximation is good. If the initial approximation is too far from the true one, then the method diverges.

A summary of the computational method is in order. The steps are as follows.

1. Produce an initial approximation to $u(x)$ on the interval $0 \leq x \leq x_1$ and store it.

2. Produce the particular and homogeneous solutions on the interval and store them (use Eqs. (4.82)-(4.85)).

3. Solve the linear algebraic equations (4.86).

4. Produce the new approximation to $u(x)$ using Eqs. (4.81).

5.   Repeat Steps 2 - 4 as needed.

Modifications of the procedure are sometimes needed
as, for example, when storage of all of complementary solutions is not possible, or when the homogeneous solutions become nearly non-independent.  Please refer to the references
at the end of this chapter.


## 4.6     NUMERICAL METHOD AND RESULTS FOR INVERSE PROBLEMS

We apply the quasilinearization method to the inverse
problem for estimating source distributions.  We regard the
unknown constants  a  and  b  in Eq. (4.68) as functions of
x  which satisfy the differential equations,

$$a'(x) = 0 , \qquad b'(x) = 0 . \tag{4.87}$$

Their initial values are to be determined.  Since the functions  X  and  Y  do not depend upon  a  or  b,  they are
known to us numerically as solutions of their initial value
problems.  The nine equations to which we apply quasilinearization are

$$e_i' = - v_i^{-1} e_i + v_i^{-1} \frac{\lambda}{2} X_i \sum_{k=1}^{M} e_k w_k$$

$$+ v_i^{-1} X_i \, 0.5[\tanh 10(x-a) - \tanh 10(x-b)] ,$$

$$i = 1, 2, \ldots, 7 ,$$

$$a' = 0 ,$$

$$b' = 0 . \tag{4.88}$$

The linearized equations for a new approximation to $e_i(x)$, a, and b in terms of the previous approximations, $e_i^o(x)$, $a^o$, and $b^o$, are (new approximations denoted without superscript)

$$e_i' = - v_i^{-1} e_i^o + v_i^{-1} \frac{\lambda}{2} X_i \sum_{k=1}^{7} e_k^o w_k$$

$$+ v_i^{-1} X_i \ 0.5 \ [\tanh 10(x-a^o) - \tanh 10(x-b^o)]$$

$$- v_i^{-1} (e_i - e_i^o)$$

$$+ v_i^{-1} \frac{\lambda}{2} X_i \sum_{k=1}^{7} (e_k - e_k^o) w_k$$

$$+ v_i^{-1} X_i \ 0.5 [-\text{sech}^2 \ 10(x-a^o) \cdot 10(a-a^o)]$$

$$+ v_i^{-1} X_i \ 0.5 [\text{sech}^2 \ 10(x-b^o) \cdot (b-b^o)] \ ,$$

$$i = 1,2,\ldots,7 \ .$$

$$a' = 0 \ ,$$

$$b' = 0 \ . \hspace{3cm} (4.89)$$

Let us define the vector solution $u(x)$ on the interval $0 \le x \le 1.5$ whose components are as follows,

$$u_i(x) = \begin{cases} e_i(x) \ , & i = 1, 2, \ldots, 7 \ , \\ a(x) \ , & i = 8 \ , \\ b(x) \ , & i = 9 \ . \end{cases} \hspace{1cm} (4.90)$$

We express an approximation to $u(x)$ in terms of complementary solutions,

$$u_i(x) = p_i(x) + \sum_{j=1}^{2} \alpha_j h_{ij}(x) , \quad i = 1, 2, \ldots, 9 .$$

$$(4.91)$$

These solutions satisfy the equations

$$p_i' = - v_i^{-1} e_i^o + v_i^{-1} \frac{\lambda}{2} X_i \sum_{k=1}^{7} e_k^o w_k$$

$$+ v_i^{-1} X_i 0.5 [\tanh 10(x-a^o) - \tanh 10(x-b^o)]$$

$$- v_i^{-1} (p_i - e_i)$$

$$+ v_i^{-1} \frac{\lambda}{2} X_i \sum_{k=1}^{7} (p_k - e_k^o) w_k$$

$$+ v_i^{-1} X_i 0.5 [-\text{sech}^2 10(x-0^o) \cdot (p_8 - a^o)]$$

$$+ v_i^{-1} X_i 0.5 [\text{sech}^2 10(x-b^o) \cdot (p_9 - b^o)] ,$$

$$i = 1,2,\ldots,7 ,$$

$$p_8' = 0 ,$$

$$p_9' = 0 ,$$

$$(4.92)$$

$$p_i(0) = 0 , \qquad i = 1, 2, \ldots, 9 , \qquad (4.93)$$

$$h_{ij}' = -v_i^{-1} h_{ij} + v_i^{-1} \frac{\lambda}{2} X_i \sum_{k=1}^{7} h_{kj} \cdot w_k$$

$$+ v_i^{-1} X_i 0.5 [-\text{sech}^3 10(x-a^o) \cdot (10 h_{8j}]$$

$$+ v_i X_i 0.5 [\text{sech}^2 10(x-b^o) \cdot 10 h_{9j}] , \quad (4.94)$$

$$i = 1,2,\ldots,9; \ j = 1,2 .$$

$$\left[h_{ij}(0)\right] = \begin{bmatrix} 0 & 0 \\ 0 & 0 \\ 0 & 0 \\ 0 & 0 \\ 0 & 0 \\ 0 & 0 \\ 0 & 0 \\ 1 & 0 \\ 0 & 1 \end{bmatrix}, \qquad (4.95)$$

$$i = 1, 2, \ldots, 9; \quad j = 1, 2$$

These complementary solutions are produced on the interval $0 \le x \le 1.5$. The two multipliers are chosen so as to minimize $Q$, where

$$Q = \sum_{i=1}^{7} \left\{ p_i(1.5) - \sum_{j=1}^{2} \alpha_j \, h_{ij}(1.5) - \beta_i \right\}^2 . \qquad (4.96)$$

(This least squares fit requires a slight modification to the quasilinearization procedure discussed in the preceding section.) As usual, the requirement is that

$$\partial Q / \partial \alpha_1, = 0 , \qquad \partial Q / \partial \alpha_2, = 0 , \qquad (4.97)$$

or

$$\sum_{j=1}^{2} \alpha_j (\sum_i h_{ij} h_{ik}) = \sum_i (\beta_i - p_i) h_{ik}, \quad k = 1,2 .$$

$$(4.98)$$

It is easy to solve for $\alpha_1$ and $\alpha_2$ explicitly in this
case. In more general cases, a computer program for the
solution of a system of linear algebraic equations is needed.

We perform a number of computational experiments.
Let the true values of the constants be  a = 0.4,  b = 0.7.
The source distribution function and the emergent intensity
pattern for slab thickness  x = 1.5  have been shown in
Figs. 4.2 and 4.3  Beginning with the initial approximations
$a^o$ = 0.4,  $b^o$ = 0.7,  and using seven accurate values of the
emergent intensity, we take three cycles of the quasilinearization procedure and arrive at the slightly modified
estimates,  a = 0.398,  b = 0.699.  This is to be expected.
Then, starting a new experiment with moderately poor initial
guesses for the constants, we quickly converge to approximately the correct values.

In the next three experiments, errors of varying
amounts are introduced into the observations.  Seven random
numbers with a Gaussian distribution with 0.0 mean and
standard deviation 1.0 are generated.  These numbers are
multiplied by 0.001 (0.1%), 0.01 (1%), and 0.02 (2%) times
the correct intensities, respectively, in the three runs.
The products are added to the accurate measurements to produce noisy measurements.  Four iterations are carried out
per run.  The accuracies of the estimates of  a  and  b  are
presented in Table 4.1.  As a spot check, the estimated
strength of the sources at altitude 0.5 is compared with the
correct value, and the error is given in the table.  The last
column of the table lists the sums of squares of deviations, Q.

Table 4.1

Some Numerical Results

| % Error in Observations | % Error in a | % Error in b | % Error in g(0.5) | Sum Q |
|---|---|---|---|---|
| 0.0 | -0.4 | -0.14 | +0.35 | $0.2 \times 10^{-6}$ |
| 0.1 | -1.7 | -0.69 | +1.4 | $0.1 \times 10^{-5}$ |
| 1.0 | -14. | -5.8 | +6.4 | $0.8 \times 10^{-4}$ |
| 2.0 | -29.5 | -12. | +3.9 | $0.3 \times 10^{-3}$ |

4.7   RESOLVENT

The integral equation of radiative transfer in a slab containing a distribution of emitting sources is

$$u(t,x) = g(t) + \frac{\lambda}{2} \int_0^x E_1(|t-y|) \, u(y,x) \, dy , \qquad (4.100)$$
$$0 \le t \le x \le x_1 ,$$

where

$$E_1(|t-y|) = \int_0^1 e^{-(t-y)/v} \, dv/v . \qquad (4.101)$$

In terms of the resolvent, $K(t,y,x)$, the solution of Eq. (4.100) may be represented as

$$u(t,x) = g(t) + \int_0^x K(t,y,x) \, g(y) \, dy . \qquad (4.102)$$

The resolvent satisfies a Cauchy system, as we shall see.
Numerical results will be presented.

It is perhaps easier to derive the initial value
problem for the resolvent analytically rather than physically.
From the above equations we see that the resolvent satisfies
the integral equation

$$K(t,y,x) = \frac{\lambda}{2} E_1(y-t) + \frac{\lambda}{2} \int_0^x E_1(|t-y'|) K(y',y,x) \, dy',$$

$$0 \leq t, y \leq x \leq x_1 . \qquad (4.103)$$

Let us differentiate with respect to $x$ on both sides.
We obtain the integral equation,

$$K_x(t,y,x) = \frac{\lambda}{2} E_1(x-t) K(x,y,x)$$

$$+ \frac{\lambda}{2} \int_0^x E_1(|t-y'|) K_x(y',y,x) \, dy' . \qquad (4.104)$$

Using the superposition principle, we write the solution of
this integral equation as

$$K_x(t,y,x) = K(t,x,x) K(x,y,x) . \qquad (4.105)$$

Since the resolvent is symmetric, we can write

$$K(t,x,x) = K(x,t,x) = \Phi(t,x) = 2 \int_0^1 J(t,x,v) \, dv/v . \qquad (4.106)$$

Then the differential equation for $K$ becomes

$$K_x(t,y,x) = 4 \int_0^1 J(t,x,v) \; dv/v \int_0^1 J(y,x,z) \; dz/z \; .$$
$$(4.107)$$

Let us assume that $y > t$ , so that the above equation is valid for $x \geq y$ . Then the initial condition at $x = y$ is

$$K(t,y,y) = 2 \int_0^1 J(t,y,v) \; dv/v \; . \qquad (4.108)$$

The exact Cauchy system for $K(t,y,x)$ involves the source functions $J(t,x,v)$, $J(y,x,v)$, and the $X$ and $Y$ functions:

$$X_x(x,v) = \frac{\lambda}{2} Y(x,v) \int_0^1 Y(x,z) \; dz/z \; ,$$

$$Y_x(x,v) = -\frac{1}{v} Y(x,v) + \frac{\lambda}{2} X(x,v) \int_0^1 Y(x,z) \; dz/z \; ,$$

$$J_x(t,x,v) = -\frac{1}{v} J(t,x,v) + \frac{\lambda}{2} X(x,v) \int_0^1 J(t,x,z) \; \frac{dz}{z} \; ,$$

$$J_x(y,x,v) = -\frac{1}{v} J(y,x,v) + \frac{\lambda}{2} X(x,v) \int_0^1 J(y,x,z) \; \frac{dz}{z} .$$

$$K_x(t,y,x) = 4 \int_0^1 J(t,x,v) \; \frac{dv}{v} \int_0^1 J(y,x,z) \; \frac{dz}{z} \; ,$$
$$(4.109)$$

$$X(0,v) = 1 ,$$

$$Y(0,v) = 1 ,$$

$$J(t,t,v) = \frac{\lambda}{4} X(t,v) ,$$

$$J(y,y,v) = \frac{\lambda}{4} X(y,v) ,$$

$$K(t,y,y) = 2 \int_{0}^{1} J(t,y,z) \frac{dz}{z}. \tag{4.110}$$

In the numerical procedure, the integrals in the Cauchy system are approximated by sums according to a quadrature formula of order N.

Writing as usual

$$X_i(x) = X(x,v_i),$$

$$Y_i(x) = Y(x,v_i) ,$$

$$J_i(t,x) = J(t,x,v_i) ,$$

$$J_i(y,x) = J(y,x,v_i) , \qquad i = 1, 2, \ldots, N ,$$

$$\tag{4.111}$$

we have the system of ordinary differential equations,

$$X_i'(x) = \frac{\lambda}{2} Y_i(x) \sum_{j=1}^{N} Y_j(x) \frac{w_j}{v_j} , \tag{4.112}$$

$$Y_i'(x) = -\frac{1}{v_i} Y_i(x) + \frac{\lambda}{2} X_i(x) \sum_{j=1}^{N} Y_j(x) \frac{w_j}{v_j} , \tag{4.113}$$

$$J_i'(t,x) = -\frac{1}{v_i} J_i(t,x) + \frac{\lambda}{2} X_i(x) \sum_{j=1}^{N} J_j(t,x) \frac{w_j}{v_j} ,$$

$$\text{(4.114)}$$

$$J_i'(y,x) = -\frac{1}{v_i} J_i(y,x) + \frac{\lambda}{2} X_i(x) \sum_{j=1}^{N} J_j(y,x) \frac{w_j}{v_j} ,$$

$$\text{(4.115)}$$

$$K'(t,y,x) = 4 \sum_{j=1}^{N} J_j(t,x) \frac{w_j}{v_j} \sum_{j=1}^{N} J_j(y,x) \frac{w_j}{v_j} ,$$

$$\text{(4.116)}$$

and the initial conditions

$$X_i(0) = 1 , \tag{4.117}$$

$$Y_i(0) = 1 , \tag{4.118}$$

$$J_i(t,t) = \frac{\lambda}{4} X_i(t) , \tag{4.119}$$

$$J_i(y,y) = \frac{\lambda}{4} X_i(y) , \tag{4.120}$$

$$K(t,y,y) = 2 \sum_{j=1}^{N} J_j(t,y) \frac{w_j}{v_j} , \tag{4.121}$$

$$i = 1, 2, \ldots, N .$$

The 2N equations (4.112) and (4.113) are integrated from $x = 0$ to $x = t$ using the conditions in Eqs. (4.117) and (4.118). At $x = t$, the initial conditions in Eq. (4.119) are employed and the N differential Eqs. (4.114) are adjoined so that there are 3N equations on the interval $x = t$ to $x = y$. Then Eqs. (4.120) and (4.121) are used at $x = y$,

and differential Eqs. (4.115) and (4.116) are adjoined.  The
system of  4N+1  differential equations are integrated until
x  has taken on the maximum desired value.  A family of
resolvents is thus produced, and comprises a useful parameter study.

    We present some numerical results for the resolvent
using a Gaussian quadrature formula.  To integrate the ordinary
differential equations, an Adams-Moulton method with a Runge-
Kutta start was employed.  The calculations were done on an
IBM System 360, Model 44.  The  X,  Y,  J  and  $\Phi$  functions
produced in the course of the calculations compare favorably
with those computed earlier.

    In Table 4.2 are displayed values of  $K(t,y,x)$  for
the case  $\lambda = 1$, $x = 1$,  for  $t = 0, 0.2, \ldots, 0.8$  and
$y = 0.2, 0.4, \ldots, 1.0$.  The entries are paired corresponding
to step sizes of integration of  $h = 0.01$  (upper entries)
and 0.005 (lower entries).  We used  $N = 7$  in the quadrature
formula.  There is agreement to one part in $10^4$.

    To obtain the missing entries in Table 4.2, use the
symmetry relation  $K(y,t,x) = K(t,y,x)$.  The resolvent is
infinite on the line  $y = t$.

    The resolvent is also symmetric about the line
$y = x - t$,  i.e.,

$$K(t,y,x) = K(x-y, x-t, x) .$$

This is an important internal check on the calculations.  Consider, for example, the calculation of  $K(0.2, 0.4, 1)$  in
Table 4.2 as the solution of the differential equation

Table 4.2

Values of  $K(t,y,x)$  for  $\lambda = 1.0$, $x = 1.0$ Calculated
With  $N = 7$; $h = 0.005$ and $0.01$

|   |     | 0 | 0.2 | 0.4 | 0.6 | 0.8 | 1.0 |
|---|-----|---|-----|-----|-----|-----|-----|
|   | 1.0 | 0.43945 | 0.66008 | 0.85924 | 1.0837 | 1.3984 | - |
|   |     | 0.43944 | 0.66004 | 0.85919 | 1.0837 | 1.3983 | - |
|   | 0.8 | 0.66010 | 1.0026 | 1.3332 | 1.7683 | - | - |
|   |     | 0.66007 | 1.0025 | 1.3331 | 1.7683 | - | - |
| y | 0.6 | 0.85926 | 1.3332 | 1.8596 | - | - | - |
|   |     | 0.85923 | 1.3331 | 1.8595 | - | - | - |
|   | 0.4 | 1.0838 | 1.7684 | - | - | - | - |
|   |     | 1.0837 | 1.7683 | - | - | - | - |
|   | 0.2 | 1.3984 | - | - | - | - | - |
|   |     | 1.3983 | - | - | - | - | - |
|   | 0   | 0 | - | - | - | - | - |

$$K_x(0.2,0.4,x) = \Phi(0.2,x) \ \Phi(0.4,x) \ ,$$
$$0.4 < x < 1 \ ,$$

$$K(0.2,0.4,0.4) = \Phi(0.2,0.4) \ .$$

On the other hand,  $K(0.6,0.8,1)$  is computed as the solution
of

$$K_x(0.6,0.8,x) = \Phi(0.6,x) \ \Phi(0.8,x) \ ,$$
$$0.8 < x < 1 \ ,$$

$$K(0.6,0.8,0.8) = \Phi(0.6,0.8).$$

Table 4.2 for step size 0.005, shows that

$$K(0.2, 0.4, 1) = K(0.6, 0.8, 1)$$

$$= 1.7683 .$$

Table 4.3 shows values of the resolvent for the case $\lambda = 1$, $x = 2$, calculated with $N = 7$ (upper entries) and $N = 5$ (lower entries) quadrature points, and a step size of 0.01. Maximum discrepancy here is about 1 part in $10^3$.

In Tables 4.4 - 4.6, the resolvent is tabulated for the cases $\lambda = 1$, and $x = 5$; $\lambda = 0.5$ and $x = 2$; and $\lambda = 0.1$ and $x = 2$, respectively. The calculations were performed with $N = 7$ and step size 0.01. Computing time was in the order of a few minutes, no special care having been taken to make the program as efficient as possible.

The calculations described generate finite values for the function $K(t, y, x)$ along the line $y = t$, where in fact it has a logarithmic singularity. As the order of the quadrature formula increases, the calculated values necessarily diverge here.

The calculated values of $K(t, y, x)$ were used to generate a solution to the equation

$$u(t) = g(t) + \frac{\lambda}{2} \int_0^x E_1(|t-y|) \, u(y) \, dy \qquad (4.122)$$

by the formula

$$u(t) = g(t) + \int_0^x K(t, y, x) \, g(y) \, dy . \qquad (4.123)$$

Table 4.3

Values of   K(t,y,x)   for   $\lambda$ = 1.0, x = 2.0
Calculated with   N =  5  and   7; h = 0.01

|   |   | 0 | 0.4 | 0.8 | 1.2 | 1.6 | 2.0 |
|---|---|---|-----|-----|-----|-----|-----|
|   | 2.0 | 0.296 | 0.555 | 0.783 | 1.023 | 1.330 | - |
|   |     | 0.296 | 0.555 | 0.782 | 1.024 | 1.330 | - |
|   | 1.6 | 0.555 | 1.046 | 1.493 | 2.025 | - | - |
|   |     | 0.555 | 1.045 | 1.493 | 2.025 | - | - |
| y | 1.2 | 0.783 | 1.493 | 2.204 | - | - | - |
|   |     | 0.782 | 1.493 | 2.205 | - | - | - |
|   | 0.8 | 1.023 | 2.025 | - | - | - | - |
|   |     | 1.024 | 2.025 | - | - | - | - |
|   | 0-4 | 1.330 | - | - | - | - | - |
|   |     | 1.330 | - | - | - | - | - |
|   | 0 | - | - | - | - | - | - |

Table 4.4

Values of  K(t,y,x) for  $\lambda$ = 1.0, x = 5.0

|   |     | 0.0   | 1.0   | 2.0   | 3.0   | 4.0   | 5.0 |
|---|-----|-------|-------|-------|-------|-------|-----|
|   | 5.0 | 0.156 | 0.458 | 0.731 | 1.007 | 1.308 | -   |
|   | 4.0 | 0.458 | 1.349 | 2.160 | 3.012 | -     | -   |
| y | 3.0 | 0.731 | 2.160 | 3.497 | -     | -     | -   |
|   | 2.0 | 1.007 | 3.012 | -     | -     | -     | -   |
|   | 1.0 | 1.308 | -     | -     | -     | -     | -   |
|   | 0.0 | -     | -     | -     | -     | -     | -   |

Table 4.5

Values of  K(t,y,x) for  $\lambda$ = 0.5, x = 2.0

|   |     | 0      | 0.4    | 0.8    | 1.2   | 1.6   | 2.0 |
|---|-----|--------|--------|--------|-------|-------|-----|
|   | 2.0 | 0.0292 | 0.0538 | 0.0897 | 0.152 | 0.283 | -   |
|   | 1.6 | 0.0538 | 0.101  | 0.175  | 0.326 | -     | -   |
| y | 1.2 | 0.0897 | 0.175  | 0.333  | -     | -     | -   |
|   | 0.8 | 0.152  | 0.326  | -      | -     | -     | -   |
|   | 0.4 | 0.283  | -      | -      | -     | -     | -   |
|   | 0   | -      | -      | -      | -     | -     | -   |

MULTIPLE SCATTERING

Table 4.6

Values of  K(t,y,x) for  $\lambda = 0.1$, x = 2.0

|        | 0.0     | 0.4     | 0.8     | 1.2    | 1.6    | 2.0 |
|--------|---------|---------|---------|--------|--------|-----|
| 2.0    | 0.00283 | 0.00503 | 0.00910 | 0.0174 | 0.0382 | -   |
| 1.6    | 0.00503 | 0.00930 | 0.0178  | 0.0391 | -      | -   |
| y  1.2 | 0.00910 | 0.0178  | 0.0392  | -      | -      | -   |
| 0.8    | 0.0174  | 0.0391  | -       | -      | -      | -   |
| 0.4    | 0.0382  | -       | -       | -      | -      | -   |
| 0      | -       | -       | -       | -      | -      | -   |

To avoid the logarithmic singularity in the integrand of
Eq. (4.123), the expression for evaluating  u(t)  was rewritten as

$$u(t) = g(t) + \int_0^x \left[ K(t,y,x) + \frac{\lambda}{2} \log|t-y| \right] g(y) \; dy$$

$$- \frac{\lambda}{2} \int_0^x \log|t-y| \; [g(y) - g(t)] \; dy$$

$$- \frac{\lambda}{2} g(t) \int_0^x \log|t-y| dy \; . \qquad (4.124)$$

The first two integrals were evaluated using Simpson's rule
for numerical quadrature, and the last integral was evaluated
analytically.

In the case  $g(t) = (\lambda/4)\exp(-|x-t|/v)$, where
$\lambda = 1$, x = 1, v = 0.5  and  $0 \leq t \leq 1.0$,  the calculated values

of u(t) should be those of the corresponding values of
J(t,x,v). This provides a consistency check on the values
obtained for K(t,y,x). For the values of K(t,y,x),
see Table 4.2. Values of K(t,y,x) + $(\lambda/2)\log|t-y|$ along
the line t = y were obtained by extrapolating calculated
values of this expression for fixed t and all available
values of y ≠ t using third order differences. Table 4.7
gives the values found for this expression. Table 4.8 com-
pares the values of u(t) evaluated as described with known
values for J(t,x,v). The values evidently agree to within
about 1 percent.


4.8     REPRESENTATION FORMULA

        The functions $\Phi(t,x)$ and u(t,x) satisfy the
integral equations

$$\Phi(t,x) = \frac{\lambda}{2} E_1(x-t) + \frac{\lambda}{2} \int_0^x E_1(|t-y|) \, \Phi(y,x) \, dy \, ,$$
$$(4.125)$$

$$u(t,x) = g(t) + \frac{\lambda}{2} \int_0^x E_1(|t-y|) \, u(y,x) \, dy \, , \qquad (4.126)$$

$$0 \le t \le x \le x_1 \, .$$

We will derive a formula which expresses u(t,x) in terms
of $\Phi(t,x)$.
        Let us first derive a relation which we will need
later on. Consider the two integral equations Eq. (4.125)
and (4.126). Cross multiply terms and integrate as follows:

Table 4.7

The Function $K(t,y,x) + (\lambda/2)\log|t-y|$ for the Case $\lambda = 1$, $x = 1$

| y | 0 | 0.1 | 0.2 | 0.3 | 0.4 | 0.5 | 0.6 | 0.7 | 0.8 | 0.9 | 1.0 |
|---|---|---|---|---|---|---|---|---|---|---|---|
| 1.0 | 0.4394 | 0.5070 | 0.5485 | 0.5802 | 0.6038 | 0.6192 | 0.6256 | 0.6201 | 0.5936 | 0.5188 | 0.3177 |
| 0.9 | 0.5070 | 0.6027 | 0.6664 | 0.7190 | 0.7628 | 0.7988 | 0.8285 | 0.8378 | 0.8146 | 0.6917 | 0.5188 |
| 0.8 | 0.5485 | 0.6664 | 0.7471 | 0.8156 | 0.8750 | 0.9258 | 0.9636 | 0.9700 | 0.8742 | 0.8146 | 0.5936 |
| 0.7 | 0.5803 | 0.7190 | 0.8156 | 0.8990 | 0.9724 | 1.0326 | 1.0631 | 0.9860 | 0.9700 | 0.8378 | 0.6201 |
| 0.6 | 0.6038 | 0.7628 | 0.8750 | 0.9724 | 1.0548 | 1.1072 | 1.6101 | 1.0631 | 0.9636 | 0.8258 | 0.6256 |
| 0.5 | 0.6192 | 0.7988 | 0.9258 | 1.0326 | 1.1072 | 1.1219 | 1.1022 | 1.0326 | 0.9258 | 0.7987 | 0.6192 |
| 0.4 | 0.6256 | 0.8258 | 0.9636 | 1.0631 | 0.6101 | 1.1072 | 1.0548 | 0.9724 | 0.8250 | 0.7628 | 0.6038 |
| 0.3 | 0.6201 | 0.8378 | 0.9701 | 0.9860 | 1.0631 | 1.0326 | 0.9724 | 0.8990 | 0.8156 | 0.7190 | 0.5802 |
| 0.2 | 0.5936 | 0.8146 | 0.8742 | 0.9701 | 0.9636 | 0.9258 | 0.8750 | 0.8156 | 0.7471 | 0.6664 | 0.5485 |
| 0.1 | 0.5189 | 0.6917 | 0.8146 | 0.8378 | 0.8258 | 0.7988 | 0.7628 | 0.7190 | 0.6664 | 0.6027 | 0.5070 |
| 0 | 0.3177 | 0.5189 | 0.5936 | 0.6201 | 0.6256 | 0.6192 | 0.6038 | 0.5803 | 0.5485 | 0.5070 | 0.4394 |

Table 4.8

Comparison Between  u(t)  and  J(t,x,v)  for the Case  $\lambda = 1$, $x = 1$, $v = 0.5$

| t | 0 | 0.1 | 0.2 | 0.3 | 0.4 | 0.5 | 0.6 | 0.7 | 0.8 | 0.9 | 1.0 |
|---|---|-----|-----|-----|-----|-----|-----|-----|-----|-----|-----|
| u(t) | 0.4991 | 0.6396 | 0.7600 | 0.8763 | 0.9860 | 1.114 | 1.219 | 1.348 | 1.462 | 1.554 | 1.571 |
| J(t,1,0.5) | 0.5000 | 0.6423 | 0.7628 | 0.8809 | 0.9991 | 1.119 | 1.239 | 1.358 | 1.471 | 1.566 | 1.574 |

$$\Phi(t,x) \left[ g(t) + \frac{\lambda}{2} \int_0^x E_1(|t-y|) \ u(y,x) \ dy \right] dt$$

$$= \int_0^x u(t,x) \left[ \frac{\lambda}{2} E_1(x-t) + \frac{\lambda}{2} \int_0^x E_1(|t-y|) \Phi(y,x) \ dy \right] dt.$$

$$(4.127)$$

After an interchange in the order of integration, this results in the relation (Hopf Lemma),

$$\int_0^x \Phi(t,x) \ g(t) \ dt = \frac{\lambda}{2} \int_0^x u(t,x) \ E_1(x-t) \ dt .$$

$$(4.128)$$

We use this formula to express $u(x,x)$ in terms of $\Phi(t,x)$,

$$u(x,x) = g(x) + \frac{\lambda}{2} \int_0^x E_1(x-y) \ u(y,x) \ dy$$

$$= g(x) + \int_0^x \Phi(y,x) \ g(y) \ dy . \qquad (4.129)$$

Recall the differential equation satisfied by $u(t,x)$,

$$u_x(t,x) = u(x,x) \ \Phi(t,x) . \qquad (4.130)$$

For an initial condition let us state, from Eq. (4.129),

$$u(t,t) = g(t) + \int_0^x \Phi(y,t) \ g(y) \ dy . \qquad (4.131)$$

Now the analytical solution of this Cauchy problem is

$$u(t,x) = u(t,t) + \int_t^x u(x',x') \; \Phi(t,x') \; dx' . \qquad (4.132)$$

Introduce the function $U(x)$,

$$U(x) = u(x,x) , \qquad\qquad\qquad 0 \leq x \leq x_1 , \qquad (4.133)$$

Then the formula can be written,

$$u(t,x) = U(t) + \int_t^x U(x') \; \Phi(t,x') \; dx' , \qquad (4.134)$$

where , from Eqs. (4.133) and (4.131) ,

$$U(s) = g(s) + \int_0^s \Phi(y,s) \; g(y) \; dy . \qquad (4.135)$$

This formula, Eq. (4.134), may be compared to the one involving the resolvent, $K(t,y,x)$, Eq. (4.102). This formula is superior in that it requires knowledge of a function of fewer variables, $\Phi(y,s)$. Note, however, that the function $\Phi(y,s)$ must be known for all $0 \leq y \leq s$, and $t \leq s \leq x$ , or at least at enough points such that a quadrature formula might accurately be applied. Since it may be desired to evaluate the integrals in Eqs. (4.134) and (4.135) via a computer algorithm, let us point out that the numerical solution of the Cauchy system as discussed in Section 4.3 is such an algorithm, and a very effective one at that.

## EXERCISES

1. Derive analytically the Cauchy system for the internal intensity function $I(t,v,x)$ of this chapter.

2. In quasilinearization, the new approximation satisfies the linear differential system. Assuming that the new approximation may be expressed in terms of the particular and homogeneous solutions, derive the differential equations for the $\left\{p_i(x)\right\}$ and $\left\{h_{ij}(x)\right\}$.

3. How would you modify the quasilinearization method so as to have minimum storage requirements?

4. How would you overcome instability of the quasilinear differential equations, if the problem should arise?

5. Spend some time in interpreting and trying to understand the results of the numerical experiments as summarized in Table 4.1. What are your conclusions?

6. Derive the integral equation for the resolvent either physically or analytically, or both.

7. Prove that the resolvent is symmetric in its first two arguments.

8. Derive the Cauchy system for the resolvent physically.

9. Derive the equation for $u(t,x)$ in terms of the resolvent (Eq. 4.102), physically.

10. Write computer codes and reproduce the numerical results given in this chapter.

# REFERENCES

1.  R. Bellman, H. Kagiwada, R. Kalaba and S. Ueno, "Numerical Results for the Estimation of Source Distribution from External Radiation Field Measurements," *J. Computational Physics*, v. 1, no. 4, 1967, pp. 457-470.

2.  R. Bellman and R. Kalaba, *Quasilinearization and Nonlinear Boundary Value Problems*, American Elsevier Publishing Co., New York, 1965.

3.  H. Kagiwada, *System Identification: Methods and Applications*, Addison-Wesley Publishing Co., Reading, Mass, 1974.

4.  R. Bellman, H. Kagiwada and R. Kalaba, "Wengert's Numerical Method for Partial Derivatives, Orbit Determination, and Quasilinearization," *Communications of the Association for Computing Machinery*, v. 8, no. 4, 1965, pp. 231-232.

5.  J. Buell, H. Kagiwada, R. Kalaba, A. McNabb and A. Schumitzky, "Computation of the Resolvent for the Auxiliary Equation of Radiative Transfer," *J. Quant. Spectros . Radiat. Transfer*, v. 8, 1968, pp. 1481-1489.

CHAPTER 5

INHOMOGENEOUS MEDIA


5.1    DERIVATION OF CAUCHY SYSTEM

We consider the monodirectional illumination of a
plane parallel, inhomogeneous medium of optical thickness
x with albedo for single scattering $\lambda(t)$ at altitude t
in the medium, $0 \leq t \leq x \leq x_1$ . The net flux is $\pi$ and the
direction cosine of the incoming rays is z. The source
function is denoted $J(t,x,z)$ .

By the addition of a layer of thickness $\Delta$ to the top
of the medium of thickness x, enumeration of the changes in
the source function, and passage to the limit as $\Delta$ tends to 0 ,
the differential integral equation for $J(t,x,z)$ is obtained,

$$J_x(t,x,z) = - z^{-1} J(t,x,z)$$

$$+ 2 J(x,x,z) \int_0^1 J(t,x,z') \, dz'/z' . \quad (5.1)$$

This equation has the same form as that for the homogeneous

medium. A difference, though, appears in the calculation of $J(x,x,z)$.

Define the function $r(v,z,x)$ to be the intensity of the radiation diffusely reflected with direction cosine $v$, by the slab of thickness $x$, with parallel ray illumination having direction cosine $z$ and net flux $\pi$. This is the emergent intensity at the top. Diffuse radiation is that which has been scattered one or more times in the medium.

The rate of scattering at the top in a cylinder with unit base area and height $\Delta$ is expressible as

$$J(x,x,z)\Delta = \pi z \frac{\Delta}{z} \frac{\lambda(x)}{4\pi}$$

$$+ \pi \int_0^1 zr(v',z,x)\ 2\pi dv'\ v'\ \frac{\Delta}{v'} \frac{\lambda(x)}{4\pi} + o(\Delta)$$

$$= \Delta\lambda(x) \frac{1}{4} \left\{ 1 + 2 \int_0^1 r(v',z,x)\ dv' \right\} + o(\Delta). \tag{5.2}$$

The source function at the top is therefore expressible in terms of the function $r(v,z,x)$ as

$$J(x,x,z) = \frac{\lambda(x)}{4} \left\{ 1 + 2 \int_0^1 r(v',z,x)\ dv' \right\}. \tag{5.3}$$

It is appropriate to introduce the symmetric function $R(v,z,x)$,

$$r(v,z,x) = R(v,z,x)/4v. \tag{5.4}$$

Then the formula for $J(x,x,z)$ becomes

$$J(x,x,z) = \frac{\lambda(x)}{4} \left\{ 1 + \frac{1}{2} \int_0^1 R(v',z,x) \; dv'/v' \right\} \; . \quad (5.5)$$

The differential equation for $J(t,x,z)$ is written

$$J_x(t,x,z) = - z^{-1} J(t,x,z)$$

$$+ \frac{\lambda(x)}{2} \left\{ 1 + \frac{1}{2} \int_0^1 R(v',z,x) \; dv'/v' \right\}$$

$$\cdot \int_0^1 J(t,x,z') \; dz'/z',$$

$$x \geq t \; . \quad (5.6)$$

The initial condition is

$$J(t,t,z) = \frac{\lambda(t)}{4} \left\{ 1 + \frac{1}{2} \int_0^1 R(v',z,t) \; dv'/v' \right\}. \quad (5.7)$$

Next we must determine the function $R(v,z,x)$.

Consider the slabs of thicknesses $x + \Delta$ and $x$. Relating the reflected intensities of these slabs, we see that

$$r(v,z,x+\Delta) = r(v,z,x) \; [1 - (z^{-1} + v^{-1})\Delta]$$

$$+ J(x,x,z)\Delta \left\{ v^{-1} + \int_0^1 (\pi z')^{-1} 2\pi dz' r(v,z',x) \right\}$$

$$+ o(\Delta) \; . \quad (5.8)$$

The first terms shows the loss of energy due to absorption in the top layer of thickness $\Delta$. Next we account for the gain in reflected intensity due to scattering at the top.

The scattered radiation either emerges directly or it emerges
after multiple scattering.

Use is made of Eq. (5.3) to arrive at the integro-
differential equation for $r(v,z,x)$ ,

$$r_x(v,z,x) = -(z^{-1} + v^{-1})\, r(v,z,x)$$

$$+ \frac{\lambda(x)}{4} \left\{ 1 + 2 \int_0^1 r(v',z,x)\, dv' \right\}$$

$$\cdot \left\{ v^{-1} + 2 \int_0^1 r(v,z',x)\, dz'/z' \right\},$$
$$x \geq 0. \quad (5.9)$$

In the limiting case of a slab of zero thickness overlying
a complete absorber, the diffusely reflected intensity is

$$r(v,z,0) = 0 . \qquad\qquad\qquad\qquad\qquad (5.10)$$

This equation will be modified when there is a reflecting
surface.  See Chapter 6.

The Cauchy system for the function $R(v,z,x)$   is

$$R_x(v,z,x) = -(z^{-1} + v^{-1})\, R(v,z,x)$$

$$+ \lambda(x) \int_0^1 \left\{ 1 + \frac{1}{2} \int_0^1 R(v',z,x)\, dv'/v' \right\}$$

$$\cdot \left\{ 1 + \frac{1}{2} \int_0^1 R(v,z',x)\, dz'/z' \right\},$$
$$x \geq 0 , \quad (5.11)$$

$$R(v,z,0) = 0 . \qquad\qquad\qquad\qquad\qquad (5.12)$$

The Cauchy system for the source function is composed of
Eqs. (5.6), (5.7), (5.11) and (5.12).

Let us rederive the Cauchy system analytically beginning with the integral equation,

$$J(t,x,z) = \frac{\lambda(t)}{4} e^{-(x-t)/z}$$

$$+ \frac{\lambda(t)}{2} \int_0^x E_1(|t-y|) \, J(y,x,z) \, dy \, ,$$
$$0 \le t \le x \le x_1 \, . \quad (5.13)$$

Differentiate on both sides of the above equation with
respect to  x, and we obtain

$$J_x(t,x,z) = - z^{-1} \frac{\lambda(t)}{4} e^{-(x-t)/z} + \frac{\lambda(t)}{2} E_1(x-t) \, J(x,x,z)$$

$$+ \frac{\lambda(t)}{2} \int_0^x E_1(|t-y|) \, J_x(y,x,z) \, dy. \quad (5.14)$$

The solution of this integral equation for  $J_x(t,x,z)$   may
be expressed in the form

$$J_x(t,x,z) = - z^{-1} J(t,x,z)$$

$$+ J(x,x,z) \, 2 \int_0^1 J(t,x,v) \, dv/v \, . \quad (5.15)$$

The source function evaluated at  $t = x$   is

$$J(x,x,z) = \frac{\lambda(x)}{4} + \frac{\lambda(x)}{2} \int_0^x E_1(x-y) \, J(y,x,z) \, dy$$

$$= \frac{\lambda(x)}{4} \left\{ 1 + 2 \int_0^x \int_0^1 v^{-1} e^{-(x-y)/v} dv \, J(y,x,z) dy \right\}.$$
$$(5.16)$$

Introduce the function $R(v,z,x)$,

$$R(v,z,x) = 4 \int_0^x e^{-(x-y)/v} \, J(y,x,z) \, dy \; . \qquad (5.17)$$

Thus we have

$$J(x,x,z) = \frac{\lambda(x)}{4} \left\{ 1 + \frac{1}{2} \int_0^1 R(v,z,x) \, dv/v \right\}. \qquad (5.18)$$

To obtain a differential equation for $R(v,z,x)$, we differentiate on both sides of Eq. (5.17),

$$R_x(v,z,x) = 4 \, J(x,x,z) - v^{-1} \, R(v,z,x)$$

$$+ 4 \int_0^x e^{-(x-y)/v} \, J_x(y,x,z) \, dy \; . \qquad (5.19)$$

We evaluate the integral using Eq. (5.15) for $J_x$. It is

$$4 \int_0^x e^{-(x-y)/v}$$

$$\cdot \left\{ - z^{-1} \, J(y,x,z) + J(x,x,z) \, 2 \int_0^1 J(y,x,v')dv'/v' \right\} dy$$

$$= - z^{-1} \, R(v,z,x) + J(x,x,z) \, 2 \int_0^1 R(v,v',x)dv'/v'.$$
$$(5.20)$$

Then Eq. (5.19) becomes

$$R_x(v,z,x) = - (v^{-1} + z^{-1}) \, R(v,z,x)$$

$$+ 4 \, J(x,x,z) \left\{ 1 + \frac{1}{2} \int_0^1 R(v,v',x) \, dv'/v' \right\} \tag{5.21}$$

or, finally,

$$R_x(v,z,x) = - (v^{-1} + z^{-1}) \, R(v,z,x)$$

$$+ \lambda(x) \left\{ 1 + \frac{1}{2} \int_0^1 R(v',z,x) \, dv'/v' \right\}$$

$$\cdot \left\{ 1 + \frac{1}{2} \int_0^1 R(v,z',x) \, dz'/z' \right\} . \tag{5.22}$$

Evaluation of $R(v,z,0)$ follows from Eq. (5.17),

$$R(v,z,0) = 0 . \tag{5.23}$$

The initial condition on $J(t,x,z)$ follows from Eq. (5.18),

$$J(t,t,z) = \frac{\lambda(t)}{4} \left\{ 1 + \frac{1}{2} \int_0^1 R(v,z,t) \, dv/v \right\} . \tag{5.24}$$

Let us consider the intensity of diffusely transmitted light, and define $t(v,z,x)$ to be the intensity of the radiation diffusely transmitted with direction cosine $v$, by the slab of thickness $x$, with parallel ray illumination having direction cosine $z$ and net flux $\pi$. The function $T(v,z,x)$ is defined by the relation

$$t(v,z,x) = \frac{T(v,z,x)}{4v} \quad . \tag{5.25}$$

From either the physical or analytical procedure, it can be shown that $T(v,z,x)$ satisfies the differential equation

$$T_x(v,z,x) = - z^{-1} T(v,z,x)$$

$$+ \lambda(x) \left\{ 1 + \frac{1}{2} \int_0^1 R(v',z,x) \; dv'/v' \right\}$$

$$\cdot \left\{ e^{-x/v} + \frac{1}{2} \int_0^1 T(v,z',x) \; dz'/z' \right\} \; ,$$
$$0 \leq x \leq x_1 \quad . \tag{5.26}$$

and the initial condition

$$T(v,z,0) = 0 \quad . \tag{5.27}$$

## 5.2    NUMERICAL METHOD

We approximate integrals by sums according to a quadrature formula of order $N$. Let $\{z_i\}$ and $\{w_i\}$ denote the points and weights appropriate for the quadrature, for $i = 1, 2, \ldots, N$. Introduce the functions

$$R_{ij}(x) = R(z_i, z_j, x) \; ,$$

$$J_i(x) = J(t,x,z_i)$$

$$i = 1, 2, \ldots, N; \quad j = 1, 2, \ldots, N \; . \tag{5.28}$$

The Cauchy system of Eqs. (5.11) and (5.12) is approximated
by the system of ordinary differential equations,

$$R'_{ij}(x) = -(z_i^{-1} + z_j^{-1}) R_{ij}$$

$$+ \lambda(x) \left\{ 1 + 0.5 \sum_{k=1}^{N} R_{kj} w_k/z_k \right\}$$

$$\cdot \left\{ 1 + 0.5 \sum_{k=1}^{N} R_{ik} w_k/z_k \right\} , \qquad (5.29)$$

$$x \geq 0 ,$$

$$R_{ij}(0) = 0 ,$$
$$i, j = 1, 2, ..., N . \qquad (5.30)$$

The equations for computing the source function are

$$J'_i(x) = - z_i^{-1} J_i$$

$$+ 0.5 \lambda(x) \left\{ 1 + 0.5 \sum_{k=1}^{N} R_{ki} w_k/z_k \right\}$$

$$\cdot \sum_{k=1}^{N} J_k w_k/z_k , \qquad t \leq x \leq x_1 , \qquad (5.31)$$

$$J_i(t) = 0.25 \lambda(t) \left\{ 1 + 0.5 \sum_{k=1}^{N} R_{ki} w_k/z_k \right\} , \qquad (5.32)$$

$$i = 1, 2, ..., N .$$

The computational procedure is to apply initial conditions in Eq. (5.30) and integrate Eqs. (5.29) from x = 0
to x = t. Then adjoin Eqs. (5.31) with initial conditions

(5.32), and continue the integration until x has the maximum
value, $x_1$. A similar procedure is used to compute transmission functions.

Extensive computations have been made for reflection functions of inhomogeneous slabs. The ability to compute these functions via Cauchy systems has an important
bearing on the present capability of solving inverse problems.


5.3    NUMERICAL RESULTS

Consider an inhomogeneous medium composed of two
layers. The total optical thickness is 1.0, the thickness
of each slab is 0.5, and the albedos are 0.4 in the lower
layer, 0.6 in the upper layer. In order to have a continuous,
differentiable function for the albedo, we express $\lambda$ by the
function,

$$\lambda(t) = 0.5 + 0.1 \tanh 10(t-0.5)$$
$$0 \le t \le 1 . \qquad (5.33)$$

This function is a very good approximation to the layered
medium with a region of very rapid transition from one albedo
to the other in the middle of the slab. The reflected
intensities evaluated at the points of the Gaussian quadrature formula of order N = 7, $\{r_{ij}\}$, are given in
Table 5.1.

Table 5.1

Reflected Intensities for Layered Medium

| | i = 1 | 2 | 3 | 4 | 5 | 6 | 7 |
|---|---|---|---|---|---|---|---|
| j = 1 | 0.079914 | 0.028164 | 0.014304 | 0.009104 | 0.006707 | 0.005515 | 0.004970 |
| 2 | 0.143038 | 0.091522 | 0.058437 | 0.040826 | 0.031405 | 0.026378 | 0.023989 |
| 3 | 0.167000 | 0.134331 | 0.099653 | 0.075106 | 0.060044 | 0.051445 | 0.047248 |
| 4 | 0.178898 | 0.157955 | 0.126408 | 0.099392 | 0.081253 | 0.070435 | 0.065042 |
| 5 | 0.185284 | 0.170817 | 0.142072 | 0.114229 | 0.094495 | 0.082423 | 0.076332 |
| 6 | 0.188723 | 0.177733 | 0.150791 | 0.122665 | 0.102104 | 0.089349 | 0.082870 |
| 7 | 0.190354 | 0.180898 | 0.154995 | 0.126773 | 0.105829 | 0.092748 | 0.086083 |

## 5.4    INVERSE PROBLEMS

A basic problem in radiative transfer is the estimation of the physical parameters of a scattering and absorbing atmosphere based on measurements of the diffusely reflected light.  In this section, we first take up some inverse problems for a layered medium.  We show that a large number of accurate observations leads rapidly to accurate estimates of the structure of the medium.  Next, we discuss various computational experiments conducted on the estimation of a parabolic albedo function using data containing various amounts and types of noise, studying the effect of the number and quality of measurements and of the criterion for fitting.

Computer programs and additional details may be found in Ref.2.

Layered Media.  Let a layered medium be described by the function,

$$\lambda(t) = a + b \tanh 10(t-c) , \qquad 0 \le t \le T , \qquad (5.34)$$

where the constants  a, b, c and  T  are to be estimated on the basis of the 49 accurate measurements which have been presented in Table 5.1, where  T  is the total optical thickness.  Let those measurements be denoted  $\{b_{ij}\}$  for  $i = 1, 2, \ldots, 7$; $j = 1, 2, \ldots, 7$.  The least squares criterion is to minimize the expression,

$$S = \sum_{i=1}^{7} \sum_{j=1}^{7} \left\{ r_{ij}(T) - b_{ij} \right\}^2 . \qquad (5.35)$$

As we recall, the true values are

$$T = 1.0, a = 0.5, b = 0.1, c = 0.5 . \qquad (5.36)$$

These true values will serve as checks on the results of the computational experiments.

We first perform some experiments on the estimation of  c,  the level of the interface. We hold the other constants fixed at their true values. Beginning with the initial estimates  $c = 0.2$  and  $c = 0.8$ , we quickly converge to the correct values, as shown in Table 5.2. On the other hand, when we begin with the initial approximation  $c = 0.0$ , i.e., a homogeneous slab, there is no convergence; the guess is too poor. Note how the number of correct digits in Table 5.2 increases from one approximation to the next when there is convergence.

Table 5.2

Successive Approximations of  c ,

The Level of the Interface

| Approximation | Run 1 | Run 2 | Run 3 |
|---|---|---|---|
| 0 | 0.2 | 0.8 | 0.0 |
| 1 | 0.62 | 0.57 | |
| 2 | 0.5187 | 0.5024 | No |
| 3 | 0.500089 | 0.499970 | Convergence |
| 4 | 0.499990 | 0.499991 | |
| True Value | 0.5 | 0.5 | 0.5 |

Next, let us try to estimate the overall optical thickness, T. The unknown quantity T is the end point of the range of integration, i.e., $0 \leq x \leq T$. In order to have a known end point, we define a new independent variable s through the equation,

$$s\,T = x \,, \tag{5.37}$$

so that the integration interval becomes $0 \leq s \leq 1$. Make the above change of variable in the system of ordinary differential equations for $R_{ij}$, and adjoin the equation,

$$\frac{dT}{ds} = 0 \,. \tag{5.38}$$

Solve using quasilinearization. In three trials beginning with guesses $T = 0.9$, $1.5$, and $0.5$, the value of T is correctly determined which is correct to within one part in a hundred thousand.

In the next experiment, we wish to estimate the albedoes in each of the two layers,

$$\lambda_1 \cong a - b, \quad \lambda_2 \cong a + b \,, \tag{5.39}$$

as well as the thicknesses c and $T - c$, respectively. In Table 5.3 we find the successive estimates.

## Table 5.3

Successive Approximations of $\lambda_1, \lambda_2$, and $c$

| Approximation | $\lambda_1 = a-b$ | $\lambda_2 = a+b$ | $c$ |
|---|---|---|---|
| 0 | 0.51 | 0.69 | 0.4 |
| 1 | 0.4200 | 0.6052 | 0.5038 |
| 2 | 0.399929 | 0.599995 | 0.499602 |
| 3 | 0.399938 | 0.599994 | 0.499878 |
| True Value | 0.4 | 0.6 | 0.5 |

Parabolic Profiles. The following experiments deal with inhomogeneous media having the parabolic profile,

$$\lambda(t) = 0.5 + at + bt^2 , \qquad 0 \le t \le c, \quad (5.40)$$

where a and b are constants, and the optical thickness is c. Let the true value of a particular medium be

$$a = 2, b = -2, c = 1.0 . \qquad (5.41)$$

In the first trial, the initial estimates are

$$a = 2.2, b = -1.8, c = 1.5 . \qquad (5.42)$$

After four iterations of quasilinearization, the estimates are much improved,

$$a = 1.99895, b = -1.99824, c = 1.004 . \qquad (5.43)$$

In the second trial, with the same initial estimates of a and b but with c = 0.5, the solution diverges.

In the second series of experiments, we fix the
incident angle at one of the seven angles and make accurate
observations of the diffusely reflected radiation in seven
directions. Thus there are seven trials, one for each of
the seven incident directions. In each trial the initial
guesses of the constants are the true values. Due to a
possible lack of information content in using seven measurements to estimate three constants, and due to computational
roundoff and other errors, we expect that these estimates
will deteriorate over the course of several iterations of
quasilinearization. The results after four iterations are
presented in Table 5.4 These values may be compared against
those obtained in the next series of experiments.

Table 5.4

Numerical Results with Data From
Various Input Directions

| Trial | Incident Angle | a | b | c |
|-------|---------------|---------|----------|----------|
| 1 | 88.5° | 2.00231 | -2.00456 | 0.999262 |
| 2 | 82.6° | 2.00206 | -2.00351 | 0.999361 |
| 3 | 72.7° | 2.00032 | -2.00048 | 0.999933 |
| 4 | 60.0° | 2.00072 | -1.99952 | 1.00007 |
| 5 | 45.3° | 1.99899 | -1.99841 | 1.00021 |
| 6 | 29.5° | 2.00029 | -2.00040 | 0.999972 |
| 7 | 13.0° | 1.99962 | -1.99937 | 1.00009 |
| Correct Values | | 2.0 | -2.0 | 1.0 |

In the third series of experiments, we introduce two types of noise into the measurements. Errors are denoted by signed percentages. Let $\{t_i\}$ be the true measurements of r for some fixed incident direction, $i = 1, 2, \ldots, 7$. If the noise has $\pm 5\%$ equal magnitude errors, the noisy observations are

$$n_1 = (1 + .05) \, t_1 \, ,$$

$$n_2 = (1 - .05) \, t_2 \, ,$$

$$\ldots$$

$$n_7 = (1 + .05) \, t_7 \, . \qquad\qquad (5.44)$$

If the noise has a 5% Gaussian distribution of errors, noisy observations are defined to be

$$m_i = (1 + .05 \, g_i) \, t_i \, , \qquad\qquad i = 1, 2, \ldots, 7 \, , \qquad (5.45)$$

where the $g_i$ are signed Gaussian deviates.

The results of experiments with seven noisy observations at one of the seven incident angles, or 49 noisy observations, are presented in Table 5.5. The accuracy of the estimates of the three constants is directly related to the accuracy of the data. In contrast to the trials with perfect measurements, experiments using noisy data yield more accurate results when there are many observations. When there are few observations, the incident angle has a profound effect.

In the fourth series of experiments, we study the effect of the criterion for fitting the data. The minimax

Table 5.5

Numerical Results with Errors in Observations

| Incident Angle | ±1% Equal Mag. Error | | | ±2% Equal Mag. Error | | | ±5% Equal Mag. Error | | |
|---|---|---|---|---|---|---|---|---|---|
| | a | b | c | a | b | c | a | b | c |
| 88.5° | 1.89 | -1.80 | 1.05 | 1.79 | -1.64 | 1.09 | 1.5 | -1.2 | 1.3 |
| 82.6° | 1.99 | -1.96 | 1.013 | 1.975 | -1.93 | 1.027 | 1.92 | -1.79 | 1.07 |
| 72.7° | 1.96 | -1.93 | 1.016 | 1.92 | -1.85 | 1.03 | 1.78 | -1.61 | 1.09 |
| 60.0° | 1.95 | -1.91 | 1.016 | 1.89 | -1.82 | 1.03 | 1.69 | -1.51 | 1.10 |
| 45.3° | 1.94 | -1.90 | 1.016 | 1.87 | -1.79 | 1.03 | 1.65 | -1.47 | 1.10 |
| 29.5° | 1.93 | -1.89 | 1.016 | 1.86 | -1.78 | 1.03 | 1.63 | -1.44 | 1.10 |
| 13.0° | 1.93 | -1.89 | 1.016 | 1.86 | -1.78 | 1.03 | 1.62 | -1.43 | 1.10 |
| All 7 | 1.99 | -1.98 | 1.003 | 1.96 | -1.94 | 1.009 | 1.91 | -1.85 | 1.02 |

| Incident Angle | 1% Gaussian Error | | | 2% Gaussian Error | | |
|---|---|---|---|---|---|---|
| | a | b | c | a | b | c |
| 88.5° | 1.46 | -1.2 | 1.22 | --- | --- | --- |
| 82.6° | 1.64 | -1.41 | 1.15 | --- | --- | --- |
| 72.7° | 1.71 | -1.55 | 1.09 | --- | --- | --- |
| 60.0° | 1.75 | -1.63 | 1.06 | 1.54 | -1.33 | 1.13 |
| 45.3° | 1.77 | -1.67 | 1.05 | --- | --- | --- |
| 29.5° | 1.78 | -1.68 | 1.05 | --- | --- | --- |
| 13.0° | 1.79 | -1.69 | 1.04 | --- | --- | --- |
| All 7 | 1.95 | -1.93 | 1.011 | --- | --- | --- |

condition requires the minimization of the maximum absolute
deviation,

$$\min \left\{ \max \left| R_{ij}(1) - \beta_{ij} \right| \right\}. \tag{5.46}$$

This is a linear programming problem with the linear in-
equalities,

$$\pm \beta_{ij}^{-1} \left\{ p_{ij}(1) + \sum_{k=1}^{3} c^k h_{ij}^k(1) - \beta_{ij} \right\} \le \varepsilon_{ij}, \tag{5.47}$$

$$\varepsilon_{ij} \le \varepsilon,$$

where the subscripts take on the values appropriate to the
trial under consideration. The $\{p_{ij}\}$ and $\{h_{ij}\}$ denote
the particular and homogeneous solutions, and the $\{c^k\}$ are
the multipliers. A standard linear programming code is used
to determine the constants $\{c^k\}$, $\{\varepsilon_{ij}\}$ and the maximum
deviation $\varepsilon$. Two numerical experiments are carried out,
one with $\pm$ 2% equal magnitude errors in the observations,
the other with 2% Gaussian errors. The incident angle is
$60^{\circ}$. The results are given in Table 5.6. Since the minimax
criterion is appropriate for equal magnitude errors, the
estimates are highly accurate.

Table 5.6

Numerical Results Using Minimax Criterion

| Type of Errors | Approxi-mation | a | b | c | Maximum Deviation |
|---|---|---|---|---|---|
| +– 2% equal magnitude | 0 | 2.00000 | -2.0000 | 1.00000 | -- |
| | 1 | 1.99948 | -1.99961 | 1.00001 | .0200000 |
| | 2 | 1.99948 | -1.99959 | 1.00001 | .0200000 |
| | 3 | 1.99948 | -1.99960 | 1.00001 | .0200000 |
| 2% Gaussian | 0 | 2.00000 | -2.00000 | 1.00000 | -- |
| | 1 | 1.76462 | -1.67267 | 1.03357 | .0294158 |
| | 2 | 1.76265 | -1.67487 | 1.03841 | .0293736 |
| | 3 | 1.76279 | -1.67484 | 1.03852 | .0293722 |

## EXERCISES

1.  Write a computer program to calculate  J, R  and  T
    for inhomogeneous slabs.  Let  $\lambda(t)$  be given by a
    subroutine.  Reproduce the data presented for the
    layered medium and the parabolic profile.

2.  Write out the linearized equations for inverse problems
    concerning the layered medium, and all of the equations
    needed for quasilinearization method.  Compare with
    equations in Ref. 2.

3.  Write computer programs to solve layered media and
    parabolic profile inverse problems.  Refer to Appendix
    B-3 of Ref. 2 if necessary.  Perform the cited experi-
    ments and compare your results.

4.  Derive the complete initial value problem for the internal
    intensity function for an inhomogeneous slab.

## REFERENCES

1.  R. Bellman, H. Kagiwada, R. Kalaba and S. Ueno,
    "Inverse Problems in Radiative Transfer:  Layered
    Media," *Icarus*, v. 4, no. 2, 1965, pp. 119-125.

2.  H. Kagiwada, *System Identification:  Methods and
    Applications*, Addison-Wesley Publishing Co., Reading,
    Mass., 1974.

CHAPTER 6

REFLECTING SURFACES

6.1   LAMBERT LAW REFLECTOR

A classical problem in the theory of multiple scattering of radiation is the determination of the radiation field
in a plane-parallel conservative isotropically scattering
medium bounded by a Lambert's law reflector.  Uniform
parallel rays of radiation of net flux  $\pi$  are incident on
one face of a slab of finite optical thickness  x.  Their
direction cosine with respect to an inward drawn normal is
u.  Within the slab conservative isotropic scattering takes
place, and the other face is bounded by a Lambert's law
reflector with albedo A.  The source function at optical
altitude  t  above the reflector is denoted by  J.  It
represents, of course, the rate of production of scattered
radiation per unit volume per unit solid angle at the optical
altitude  t.

It is a simple matter to see that the source function
J  satisfies the Fredholm integral equation

$$J(t,x,u,A) = \frac{1}{4} e^{-(x-t)/u} + \frac{1}{2} Aue^{-x/u} E_2(t)$$

$$+ \int_0^x \left[ \frac{1}{2} E_1(|t-y|) + A E_2(t) E_2(y) \right] J(y,x,u,A) \, dy ,$$

$$0 \leq t \leq x; \quad 0 \leq A \leq 1; \quad 0 \leq u \leq 1 . \qquad (6.1)$$

As usual, the exponential integral functions $E_1$ and $E_2$ are defined by the definite integrals

$$E_1(r) = \int_0^1 e^{-r/z} \, dz/z , \qquad\qquad r > 0 , \quad (6.2)$$

$$E_2(r) = \int_0^1 e^{-r/z} \, dz , \qquad\qquad r \geq 0 , \quad (6.3)$$

Two key problems are the finding of the diffuse scattering function $R$ and the diffuse transmission function $T$. These functions are defined by the relations

$$R(v,u,x,A) = 4 \int_0^x J(y,x,u,A) \, e^{-(x-y)/v} \, dy$$

$$+ \, 4v \, I(u,x,A) \, e^{-x/v} \qquad\qquad (6.4)$$

and

$$T(v,u,x,A) = 4 \int_0^x J(y,x,u,A) \, e^{-y/v} \, dy ,$$

$$0 \leq v,u \leq 1; \quad 0 \leq A \leq 1; \quad 0 \leq x . \quad (6.5)$$

The intensity of the radiation diffusely reflected with direction cosine $v$ is $R(v,u,x,A)/4v$, and the intensity of the diffuse radiation incident on the reflector with direction cosine $v$ is $T(v,u,x,A)/4v$, $0 \leq v \leq 1$. The

auxiliary function  I,  which represents the uniform intensity of the radiation from the Lambert's law reflector,
is given by

$$I(u,x,A) = A \left[ ue^{-x/u} + 2 \int_0^x J(y,x,u,A) \, E_2(y) \, dy \right].$$

$$(6.6)$$

The functions  R  and  T  may be computed as the
solution of two different Cauchy systems in which  x,  the
optical thickness of the medium, plays the role of the
independent variable.  Purely physical reasoning may be
used to write the Cauchy systems.  In this section we provide a firm analytical basis using the integral equation
for the source function  J  and the definitions of the
functions  R, T, and  I  as functionals on  J.  We apply
the theory as described in the book, Ref. 1.

Cauchy System for the Scattering Function.  Let us
first derive the Cauchy system for the function  $R(v,u,x,A)$

$$R_x(v,u,x,A) = -\left( \frac{1}{u} + \frac{1}{v} \right) R(v,u,x,A)$$

$$+ \left[ 1 + \frac{1}{2} \int_0^1 R(v,u',x,A) \, du'/u' \right]$$

$$\cdot \left[ 1 + \frac{1}{2} \int_0^1 R(v',u,x,A) \, dv'/v' \right],$$

$$x \geq 0 , \quad (6.7)$$

$$R(v,u,0,A) = 4Auv . \qquad (6.8)$$

Through differentiation with respect to  x  of the integral
equation (6.1) for the function  J  we obtain the relation

$$J_x(t,x,u,A) = -\frac{1}{u}\left[\frac{1}{4}e^{-(x-t)/u} + \frac{1}{2}Aue^{-x/u}E_2(t)\right]$$

$$+ \left[\frac{1}{2}E_1(x-t) + A\,E_2(t)\,E_2(x)\right]\,J(x,x,u,A)$$

$$+ \int_0^x \left[\frac{1}{2}E_1(\,|t-y|\,) + A\,E_2(t)\,E_2(y)\right]$$

$$J_x(y,x,u,A)\;dy\;. \tag{6.9}$$

Equation (6.9) is regarded as an integral equation for the function $J_x$. Its solution is seen to be

$$J_x(t,x,u,A) = -\frac{1}{u}\,J(t,x,u,A)$$

$$+ 2\,J(x,x,u,A)\int_0^1 J(t,x,u',A)\;du'/u'\;,$$
$$x \geq 0\;. \tag{6.10}$$

From the integral equation in Equation (6.1) we see that

$$J(x,x,u,A) = \frac{1}{4} + \frac{1}{2}E_2(x)\,I(u,x,A)$$

$$+ \frac{1}{2}\int_0^x E_1(x-y)\,J(y,x,u,A)\;dy. \tag{6.11}$$

According to the definition of the function $R$ we have

$$\int_0^1 \left[R(v',u,x,A)/8v'\right]\,dv' = \frac{1}{2}\int_0^x J(y,x,u,A)\,E_1(x-y)\,dy$$

$$+ \frac{1}{2}\,I(u,x,A)\,E_2(x)\;. \tag{6.12}$$

It follows that

$$J(x,x,u,A) = \frac{1}{4}\left[1 + \frac{1}{2}\int_0^1 R(v',u,x,A)\ dv'/v'\right].$$
$$(6.13)$$

Finally, for the function  $I$  we may write

$$I_x(u,x,A) = A\left[- e^{-x/u} + 2J(x,x,u,A)\ E_2(x)\right.$$

$$+ 2\int_0^x \left\{-\frac{1}{u}\ J(y,x,u,A)\right.$$

$$+ 2J(x,x,u,A)\int_0^1 J(y,x,u',A)\ du'/u'\right\} E_2(y)\,dy\bigg].$$
$$(6.14)$$

This becomes the equation

$$I_x(u,x,A) = -\frac{1}{u}\ I(u,x,A)$$

$$+ 2J(x,x,u,A)\ A$$

$$\cdot\left[E_2(x) + 2\int_0^x\int_0^1 J(y,x,u',A)\ \frac{du'}{u'}\ E_2(y)\,dy\right].$$
$$(6.15)$$

Once again employing the definition of the function  $I$  in Eq. (6.6), it is seen that the function  $I$  satisfies the differential equation

$$I_x(u,x,A) = -\frac{1}{u} I(u,x,A)$$

$$+ 2J(x,x,u,A)\int_0^1 I(u',x,A)\ du'/u' \ .$$
$$x \geq 0 \ , \qquad (6.16)$$

Having disposed of these preliminaries, we may now obtain the differential equation for the scattering function  R.

Differentiation of both sides of Eq. (6.4) with respect to  x  shows that

$$R_x(v,u,x,A) = -\frac{1}{v} R(v,u,x,A) + 4J(x,x,u,A)$$

$$+ 4\int_0^x\left[-\frac{1}{u} J(y,x,u,A) + 2J(x,x,u,A)\right.$$

$$\left.\cdot \int_0^1 J(y,x,u',A)\ du'/u'\right]e^{-(x-y)/v}dy$$

$$+ 4ve^{-x/v}\left[-\frac{1}{u} I(u,x,A)\right.$$

$$\left.+ 2J(x,x,u,A)\int_0^1 I(u',x,A)\ du'/u'\right].$$
$$(6.17)$$

Through simple algebraic manipulations this becomes

$$R_x(v,u,x,A) = -\left(\frac{1}{u} + \frac{1}{v}\right) R(v,u,x,A)$$

$$+ 4\ J(x,x,u,A)\left[1\right.$$

$$+ 2\int_0^x\int_0^1 J(y,x,u',A)\ \frac{du'}{u'}\ e^{-(x-y)/v}\ dy$$

$$\left.+ 2ve^{-x/v}\int_0^1 I(u',x,A)\ du'/u'\right] \ . \quad (6.18)$$

Next we write

$$R_x(v,u,x,A) = -\left(\frac{1}{u} + \frac{1}{v}\right) R(v,u,x,A)$$

$$+ 4 \ J(x,x,u,A) \left[1 + \frac{1}{2} \int_0^1 R(v,u',x,A)du'/u'\right],$$

$$(6.19)$$

and finally

$$R_x(v,u,x,A) = -\left(\frac{1}{u} + \frac{1}{v}\right) R(v,u,x,A)$$

$$+ \left[1 + \frac{1}{2} \int_0^1 R(v',u,x,A) \ dv'/v'\right]$$

$$\cdot \left[1 + \frac{1}{2} \int_0^1 R(v,u',x,A) \ du'/u'\right].$$

$$(6.20)$$

Equation (6.20) is the desired differential equation for the function R. The initial condition at $x = 0$ follows readily from the definition. We see that

$$R(v,u,0,A) = 4vI(u,0,A)$$

$$= 4v(Au)$$

$$= 4Auv .$$

This establishes the Cauchy system used for the function $R(v,u,x,A)$.

Cauchy System for the Transmission Function. The
transmission function T is defined by the relation

$$T(v,u,x,A) = 4 \int_0^x J(y,x,u,A) \ e^{-y/v} \ dy \ ,$$
$$0 \leq v, u \leq 1; \quad 0 \leq A \leq 1; \quad 0 \leq x \ . \quad (6.22)$$

Differentiation with respect to x shows that

$$T_x(v,u,x,A) = 4J(x,x,u,A) \ e^{-x/v}$$
$$+ \ 4 \int_0^1 \left[ - \frac{1}{u} J(y,x,u,A) \right.$$
$$+ \ 2J(x,x,u,A) \int_0^1 J(y,x,u',A) du'/u' \Big] e^{-y/v} dy$$
$$= \ - \frac{1}{u} T(v,u,x,A) + 4J(x,x,u,A) \left[ e^{-x/v} \right.$$
$$+ \ 2 \int_0^x \int_0^1 J(y,x,u',A) \ e^{-y/v} \ dy \ du'/u' \Big] .$$
$$(6.23)$$

The final result is

$$T_x(v,u,x,A) = - \frac{1}{u} T(v,u,x,A)$$
$$+ \left[ 1 + \frac{1}{2} \int_0^1 R(v',u,x,A) \ dv'/v' \right]$$
$$\cdot \left[ e^{-x/v} + \frac{1}{2} \int_0^1 T(v,u',x,A) \ du'/u' \right] ,$$
$$x \geq 0 \ . \quad (6.24)$$

The initial condition at x = 0 is

$$T(v,u,0,A) = 0 \; , \qquad\qquad 0 \leq v, \; u \leq 1 \; , \qquad (6.25)$$

according to the definition in Eq. (6.22).

Relations Between the Fields in Media Bounded by a
Reflector and a Vacuum.  Let us introduce the functions  J
and  ω  as the solutions of the integral equations

$$J(t,x,u) = \frac{1}{4} e^{-(x-t)/u} + \frac{1}{2} \int_0^x E_1(|t-y|) \; J(y,x,u) \; dy$$

$$(6.26)$$

and

$$\omega(t,x) = \frac{1}{2} E_2(t) + \frac{1}{2} \int_0^x E_1(|t-y|)\omega(Y,x) \; dy \; ,$$

$$0 \leq t \leq x \; . \qquad (6.27)$$

The integral equation (6.1) may be rewritten in the form

$$J(t,x,u,A) = \frac{1}{4} e^{-(x-t)/u} + \frac{1}{2} I(u,x,A) \; E_2(t)$$

$$+ \frac{1}{2} \int_0^x E_1(|t-y|) \; J(y,x,u,A) \; dy \; . \qquad (6.28)$$

From considerations of linearity it follows that

$$J(t,x,u,A) = J(t,x,u) + I(u,x,A) \; \omega(t,x) \qquad (6.29)$$

and

$$J(t,x,u) = J(t,x,u,0) \; . \qquad\qquad (6.30)$$

This relation enables us to express the scattering and transmission functions for a slab bounded by a Lambert's law reflector in terms of those for a slab bounded by a vacuum.

For the scattering function $R(v,u,x,A)$ we may write

$$R(v,u,x,A) = 4 \int_0^x \left[ J(y,x,u) + I(u,x,A)\, \omega(y,x) \right] e^{-(x-y)/v} dy$$

$$+ 4v\, I(u,x,A)\, e^{-x/v}$$

$$= R(v,u,x) + 4\, I(u,x,A)$$

$$\cdot \left[ v\, e^{-x/v} + \int_0^x \omega(y,x)\, e^{-(x-y)/v}\, dy \right], \quad (6.31)$$

where we have introduced the usual scattering function

$$R(v,u,x) = R(v,u,x,0)$$

$$= 4 \int_0^x J(y,x,u)\, e^{-(x-y)/v}\, dy . \quad (6.32)$$

To evaluate the integral in Equation (6.31) we note that

$$\omega(t,x) = 2 \int_0^1 J(x-t,x,u')\, du' . \quad (6.33)$$

It is seen that

$$\int_0^x \omega(y,x)\, e^{-(x-y)/v}\, dy$$

$$= 2 \int_0^x \int_0^1 J(x-y,x,u')\, du'\, e^{-(x-y)/v}\, dy$$

$$= \frac{1}{2} \int_0^1 T(v,u',x)\, du' , \quad (6.34)$$

where

$$T(v,u,x) = 4 \int_0^x J(y,x,u) \ e^{-y/v} \ dy \ . \qquad (6.35)$$

Equation (6.31) becomes

$$R(v,u,x,A) = S(v,u,x) + 4I(u,x,A)$$

$$\cdot \left[ v \ e^{-x/v} + \frac{1}{2} \int_0^1 T(v,u',x) \ du' \right] \ . \qquad (6.36)$$

For the transmission function of Equation (6.22) we have

$$T(v,u,x,A) = 4 \int_0^x \left[ J(y,x,u) + I(u,x,A)\omega(y,x) \right] e^{-y/v} \ dy$$

$$= T(v,u,x)$$

$$+ 4I(u,x,A) \int_0^x \int_0^1 2J(x-y,x,u') \ e^{-y/v} \ du'dy \ ,$$
$$(6.37)$$

or

$$T(v,u,x,A) = T(v,u,x) + 2I(u,x,A) \int_0^1 R(v,u',x) \ du' \ . $$
$$(6.38)$$

In the last equation we have, of course, used the standard terminology

$$T(v,u,x) = T(v,u,x,0)$$

$$= 4 \int_0^x J(y,x,u) \ e^{-y/v} \ dy \ . \qquad (6.39)$$

Lastly, we consider the expression for $I$ as a functional on $S(v,u,x)$ and $T(v,u,x)$. From its definition we note that

$$I(u,x,A) = A \left\{ ue^{-x/u} \right.$$

$$\left. + 2 \int_0^x \left[ J(y,x,u) + I(u,x,A)\omega(y,x) \right] E_2(y) \, dy \right\}.$$

$$(6.40)$$

Solving for $I(u,x,A)$ we find

$$I(u,x,A) = A \left[ u \, e^{-x/u} + 2 \int_0^x J(y,x,u) \, E_2(y) \, dy \right]$$

$$\cdot \left[ 1 - 2A \int_0^x \omega(y,x) \, E_2(y) \, dy \right]^{-1}. \qquad (6.41)$$

To evaluate the first integral we write

$$\int_0^x J(y,x,u) \int_0^1 e^{-y/z'} \, dz' \, dy$$

$$= \int_0^1 dz' \int_0^x J(y,x,u) \, e^{-y/z'} \, dy$$

$$= \frac{1}{4} \int_0^1 T(z',u,x) \, dz'. \qquad (6.42)$$

For the second one we see that

$$\int_0^x \omega(y,x)\ E_2(y,x)\ dy$$

$$= \int_0^x 2\int_0^1 J(x-y,x,u')\ du' \int_0^1 e^{-y/z'}\ dz'\ dy$$

$$= \int_0^1 \int_0^{-1} du'\ dz'\ 2\int_0^x J(x-y,x,u')e^{-y/z'}\ dy$$

$$= \frac{1}{2}\int_0^1 \int_0^1 R(z',u',x)\ du'\ dz'\ . \qquad (6.43)$$

The final expression for $I$ becomes

$$I(u,x,A) = A\left[u\ e^{-x/u} + \frac{1}{2}\int_0^1 T(z',u,x)\ dz'\right]$$

$$\cdot\left[1 - A\int_0^1 \int_0^1 R(z',u',x)\ du'\ dz'\right]^{-1}. \qquad (6.44)$$

The functions $R(v,u,x)$ and $T(v,u,x)$ are to be determined as the solutions of a Cauchy system. One such system is obtained by putting $A = 0$ in the differential Eqs. (6.7) and (6.24) and in the initial conditions in Eqs. (6.21) and (6.25).

Numerical Results. Let the diffusely reflected and transmitted intensities when a Lambert's law reflecter is present be $r^*$ and $t^*$, respectively,

$$r^*(v,u,x,A) = R(v,u,x,A)/4v\ , \qquad (6.45)$$

$$t^*(v,u,x,A) = T(v,u,x,A)/4v\ . \qquad (6.46)$$

Define the reflected, transmitted and global fluxes,

$$p(u,x,A) = 2\pi \int_0^1 r^*(v,u,x,A) \, v \, dv \, , \qquad (6.47)$$

$$\tau(u,x,A) = 2\pi \int_0^1 t^*(v,u,x,A) \, v \, dv \, , \qquad (6.48)$$

$$\tau_g(u,x,A) = \pi e^{-x/u} + \tau(u,x,A) \, . \qquad (6.49)$$

We use a Gaussian quadrature formula of order N to approximate integrals. We compute the R and T functions for the case of no reflector. Equation (6.44) yields the intensity I(u,x,A) for arbitrary values of A, and Eqs. (6.36) and (6.38) produce the functions R and T. Finally, intensities and fluxes are computed from Eqs. (6.45) - (6.49).

The basic numerical calculation consists of producing the reflected and transmitted intensities and fluxes for surface albedos A = 0, 0.1, 0.2, ..., 1.0, optical thicknesses 0 through 100, and conservative, isotropic scattering ($\lambda = 1$). The integration step size used is 0.005 with Gaussian quadrature of order N = 7. A check calculation involving changing the step size to 0.01 and varying the order of the quadrature from N = 3 to N = 5 is run to a thickness of 1 resulting in, at most, a change of one unit in the fourth significant figure. All calculations are programmed by J. L. Casti and performed on a CDC 6600 computer using a fourth-order Adams-Moulton predictor-corrector integration scheme. Execution time for the basic calculations is about 7 min.

An objective of the computational study is to evaluate the reflected and transmitted intensities and global transmitted fluxes. These quantities are plotted in Figures 6.1 through 6.3. These figures may be compared with corresponding

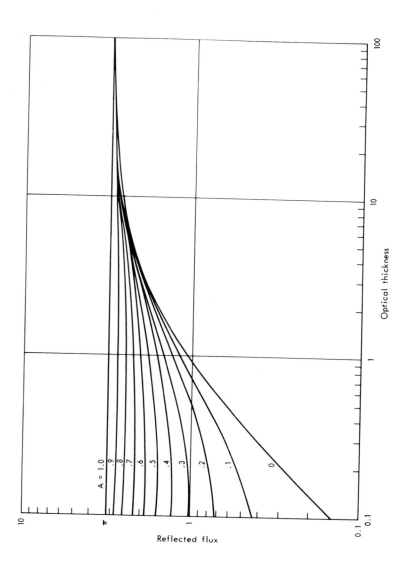

Figure 6.1   Reflected Flux at Normal Incidence for a
             Lambert Reflector with Albedo   A

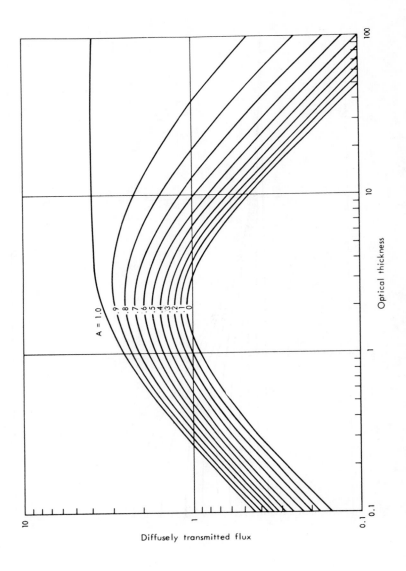

Figure 6.2   Diffusely Transmitted Flux at Normal Incidence
for a Lambert Reflector with Albedo   A

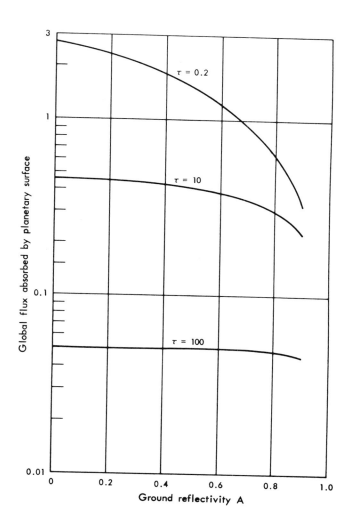

Figure 6.3    Global Flux Absorbed Versus Ground Reflectivity
for Thickness   τ = 0.2, 10, and 100, Normal
Incidence and Lambert Reflectors

Table 6.1

Reflected and Diffusely Transmitted Intensities and Fluxes for A
Slab with Conservative Isotropic Scattering Thickness 10.0,
and a Lambert Surface Albedo 0.5

REFLECTED INTENSITIES

| Inc. Angle (Deg.) | 1(88.5) | 2(82.6) | 3(72.7) | 4(60.0) | 5(45.3) | 6(29.5) | 7(13.0) | Reflected Flux |
|---|---|---|---|---|---|---|---|---|
| 1 (88.5) | .1441 | .0572 | .0339 | .0251 | .0211 | .0190 | .0181 | .0760 |
| 2 (82.6) | .2905 | .2097 | .1569 | .1286 | .1130 | .1044 | .1002 | .3820 |
| 3 (72.7) | .3953 | .3607 | .3194 | .2878 | .2666 | .2533 | .2464 | .8641 |
| 4 (60.0) | .4940 | .4974 | .4844 | .4667 | .4507 | .4388 | .4320 | 1.4277 |
| 5 (45.3) | .5824 | .6145 | .6308 | .6336 | .6296 | .6239 | .6197 | 1.9705 |
| 6 (29.5) | .6509 | .7032 | .7425 | .7642 | .7728 | .7744 | .7738 | 2.4039 |
| 7 (13.0) | .6916 | .7553 | .8084 | .8420 | .8591 | .8660 | .8679 | 2.6649 |
| Normal (00.0) | .7014 | .7678 | .8242 | .8607 | .8800 | .8882 | .8908 | 2.7281 |

TRANSMITTED INTENSITIES

| Inc. Angle (Deg.) | 1(88.5) | 2(82.6) | 3(72.7) | 4(60.0) | 5(45.3) | 6(29.5) | 7(13.0) | Transmitted Flux |
|---|---|---|---|---|---|---|---|---|
| 1 (88.5) | .0018 | .0019 | .0021 | .0023 | .0025 | .0027 | .0028 | .0078 |
| 2 (82.6) | .0112 | .0120 | .0131 | .0143 | .0155 | .0165 | .0171 | .0481 |
| 3 (72.7) | .0323 | .0345 | .0376 | .0412 | .0447 | .0475 | .0493 | .1383 |
| 4 (60.0) | .0668 | .0714 | .0778 | .0853 | .0925 | .0984 | .1020 | .2862 |
| 5 (45.3) | .1110 | .1186 | .1294 | .1417 | .1537 | .1635 | .1695 | .4757 |
| 6 (29.5) | .1548 | .1654 | .1804 | .1976 | .2144 | .2280 | .2364 | .6634 |
| 7 (13.0) | .1851 | .1978 | .2158 | .2364 | .2564 | .2727 | .2827 | .7934 |
| Normal (00.0) | .1930 | .2062 | .2249 | .2463 | .2672 | .2842 | .2947 | .8269 |

Table 6.2

Reflected and Diffusely Transmitted Intensities and Fluxes for A
Slab with Conservative Isotropic Scattering Thickness 100.0
and a Lambert Surface Albedo 1.0

REFLECTED INTENSITIES

| Inc. Angle | (Deg.) | 1(88.5) | 2(82.6) | 3(72.7) | 4(60.0) | 5(45.3) | 6(29.5) | 7(13.0) | Reflected Flux |
|---|---|---|---|---|---|---|---|---|---|
| 1 | (88.5) | .1447 | .0579 | .0347 | .0262 | .0224 | .0205 | .0196 | .0799 |
| 2 | (82.6) | .2941 | .2140 | .1623 | .1352 | .1209 | .1132 | .1096 | .4060 |
| 3 | (72.7) | .4056 | .3732 | .3350 | .3070 | .2893 | .2789 | .2737 | .9333 |
| 4 | (60.0) | .5152 | .5232 | .5167 | .5064 | .4976 | .4917 | .4885 | 1.5708 |
| 5 | (45.3) | .6176 | .6574 | .6844 | .6996 | .7076 | .7117 | .7135 | 2.2083 |
| 6 | (29.5) | .7001 | .7630 | .8174 | .8562 | .8816 | .8969 | .9046 | 2.7356 |
| 7 | (13.0) | .7505 | .8268 | .8979 | .9521 | .9892 | 1.0125 | 1.0244 | 3.0616 |
| Normal | (00.0) | .7628 | .8423 | .9175 | .9755 | 1.0156 | 1.0409 | 1.0539 | 3.1416 |

TRANSMITTED INTENSITIES

| Inc. Angle | (Deg.) | 1(88.5) | 2(82.6) | 3(72.7) | 4(60.0) | 5(45.3) | 6(29.5) | 7(13.0) | Transmitted Flux |
|---|---|---|---|---|---|---|---|---|---|
| 1 | (88.5) | .0119 | .0119 | .0119 | .0119 | .0119 | .0119 | .0119 | .0372 |
| 2 | (82.6) | .0732 | .0732 | .0732 | .0732 | .0732 | .0732 | .0732 | .2300 |
| 3. | (72.7) | .2105 | .2105 | .2105 | .2105 | .2105 | .2105 | .2105 | .6613 |
| 4 | (60.0) | .4356 | .4356 | .4356 | .4356 | .4356 | .4356 | .4356 | 1.3686 |
| 5 | (45.3) | .7239 | .7239 | .7239 | .7239 | .7239 | .7239 | .7239 | 2.2743 |
| 6 | (29.5) | 1.0096 | 1.0096 | 1.0096 | 1.0096 | 1.0096 | 1.0096 | 1.0096 | 3.1719 |
| 7 | (13.0) | 1.2077 | 1.2077 | 1.2077 | 1.2077 | 1.2077 | 1.2077 | 1.2077 | 3.7940 |
| Normal | (00.0) | 1.2587 | 1.2587 | 1.2587 | 1.2587 | 1.2587 | 1.2587 | 1.2587 | 3.9542 |

graphs in Reference 2 for the case of Rayleigh scattering.
As far as reflected and transmitted fluxes for normal
incidence are concerned, virtually no quantitative difference
exists between the reflected and transmitted fluxes for the
two scattering laws for any optical thickness in the range
0 - 100 and any surface albedo  $0 \leq A \leq 1$ .

Tables 6.1 and 6.2 present the intensities and fluxes
of the reflected and diffusely transmitted radiation for
slabs with thickness 10 and 100, and surface albedos 0.5
and 100, respectively.  These tables are excerpts from the
main calculations that print out intensities and fluxes
at 61 optical thicknesses in the range 0 - 100 for  $A = 0.0$,
0.1, 0.2, ..., 1.0.  The incident angle is constant across
a row in the tables.  It takes on one of eight different
angles as indicated.  The tables give intensities for seven
outgoing angles, and list fluxes in the last column.

An interesting property of the solution of the planetary problem with a Lambert-law reflector is obtained by
comparing this solution with that presented in a later
section for the case when the reflecting surface is a perfect
specular reflector.  It is observed for the conservative
case $\lambda = 1$, $A = 1$, that the transmitted and reflected fluxes
are virtually the same (to three figures) for all optical
thicknesses $\geq 3$.


## 6.2   LAMBERT'S LAW REFLECTERS - A REDUCTION

We reconsider the problem of determining the radiation
field produced by uniform parallel rays incident on a
finite isotropically scattering slab bounded by a Lambert's
law reflector.  We have already treated it by reduction to

a Cauchy system involving auxiliary functions of two angular
arguments.  Here the problem is reduced to a Cauchy system
involving auxiliary functions of merely one angular argument.
This is of analytic interest and computational utility.  It
leads to the hope that similar advances may be possible in
the treatment of more realistic atmospheric models.

Let uniform parallel rays of net flux $\pi$ be incident
on a finite slab of optical thickness  x.  The incident
rays make an angle arccos  u  with respect to an inward
drawn normal.  Within the slab scattering is isotropic and
conservative, and the slab is bounded on one side by a
Lambert's law reflector with albedo  A.  The source function
at optical altitude  t  (the rate of production of scattered
radiation per unit volume per unit solid angle at optical
altitude  t) is denoted by  J,

$$J = J(t,x,u,A) \ , \qquad 0 \leq t \leq x; \quad 0 \leq u \leq 1; \quad 0 \leq A \leq 1 \ . \qquad (6.50)$$

This function satisfies the Fredholm integral equation

$$J(t,x,u,A) = \frac{1}{4} e^{-(x-t)/u} + \frac{1}{2} A \ u \ e^{-x/u} E_2(t)$$

$$+ \int_0^x \frac{1}{2} E_1(|t-y|) + AE_2(t) \ E_2(y) \Big] J(y,x,u,A) \ dy \ ,$$

$$0 \leq t \leq x \ , \qquad (6.51)$$

where the functions  $E_1$  and  $E_2$  are defined by the integrals

$$E_1(r) = \int_0^1 e^{-r/z} \ dz/z \ , \qquad (6.52)$$

$$E_2(r) = \int_0^1 e^{-r/z} \, dz \, , \qquad\qquad r \geq 0 \, . \qquad (6.53)$$

Our goal is to reduce the determination of the function J to the solution of a Cauchy system involving J and several auxiliary functions of one angular argument. The independent variable of the Cauchy system will be x, the optical thickness of the medium.

Differentiation of both sides of Eq. (6.51) with respect to x yields the relation

$$J_x(t,x,u,A)$$

$$= - \frac{1}{u} \left[ \frac{1}{4} e^{-(x-t)/u} + \frac{1}{2} A \, u \, e^{-x/u} E_2(t) \right]$$

$$+ \left[ \frac{1}{2} E_1(x-t) + A E_2(t) \, E_2(x) \right] \, J(x,x,u,A)$$

$$+ \int_0^x \left[ \frac{1}{2} E_1(|t-y|) + A E_2(t) \, E_2(y) \right] J_x(y,x,u,A) \, dy.$$

$$(6.54)$$

Regard this as an integral equation for the function $J_x$, the first two terms on the right hand side being the forcing terms. Then keeping the definitions in Eqs. (6.52) and (6.53) in mind, the solution of Eq. (6.54) is seen to be

$$J_x(t,x,u,A) = - \frac{1}{u} \, J(t,x,u,A)$$

$$+ 2 \, J(x,x,u,A) \int_0^1 J(t,x,u',A) \, du'/u' \, ,$$

$$x \geq t \, . (6.55)$$

This is one of the desired differential equations.  Its
use, though, necessitates consideration of the auxiliary
function  $J(x,x,u,A)$,  the source function at  $t = x$.

Introduce the optical depth  $\tau$  into the integral
Eq. (6.51) by means of the relations

$$\tau = x-t ,$$

$$y' = x-y .$$                                              (6.56)

The result is

$$J(x-\tau,x,u,A) = \frac{1}{4} e^{-\tau/u} + \frac{1}{2} I(u,x,A) \, E_2(x-\tau)$$

$$+ \frac{1}{2} \int_0^1 E_1(|\tau-y'|) \, J(x-y',x,u,A) \, dy' ,$$
(6.57)

where  I,  the intensity of radiation leaving the Lambert's
law surface, is

$$I(u,x,A) = A \left[ ue^{-x/u} + 2 \int_0^x J(y,x,u,A) \, E_2(y) \, dy \right].$$
(6.58)

Differentiation of Eq. (6.57) with respect to  x  shows
that

$$\frac{d}{dx} J(x-\tau,x,u,A)$$

$$= \frac{1}{2} I_x(u,x,A) \ E_2(x-\tau)$$

$$+ \frac{1}{2} I(u,x,A) \ \frac{d}{dx} E_2(x-\tau)$$

$$+ \frac{1}{2} E_1(x-\tau) \ J(0,x,u,A)$$

$$+ \frac{1}{2} \int_0^x E_1(|\tau-y'|) \ \frac{d}{dx} J(x-y',x,u,A) \ dy' \ . \quad (6.59)$$

Regarding this as an integral equation for the function $(d/dx) \ J(x-\tau,x,u,A)$ and recalling that

$$\frac{d}{dx} E_2(x) = - E_1(x), \qquad\qquad x>0 \ , \qquad (6.60)$$

we see that

$$\frac{d}{dx} J(x-\tau,x,u,A)$$

$$= \left[ J(0,x,u,A) \ -I(u,x,A) \right] \ 2 \int_0^1 J(\tau,x,u') \ du'/u'$$

$$+ I_x(u,x,A) \ 2 \int_0^1 J(\tau,x,u') \ du' \ . \qquad (6.61)$$

The function $J(t,x,u)$ is defined by the relation

$$J(t,x,u) = J(t,x,u,0) \ , \qquad 0 \leq t \leq x; \ \ 0 \leq u \leq 1 \ , \quad (6.62)$$

and the function $J(t,x,u)$ itself satisfies the integral equation

$$J(t,x,u) = \frac{1}{4} e^{-(x-t)/u}$$

$$+ \frac{1}{2} \int_0^x E_1(|t-y|) \, J(y,x,u) \, dy \, , \qquad (6.63)$$

which is a special case of Eq. (6.51). Putting $\tau = 0$ yields the result ,

$$\frac{d}{dx} J(x,x,u,A)$$

$$= \left[ J(0,x,u,A) - I(u,x,A) \right] 2 \int_0^1 J(0,x,u') \, du'/u'$$

$$+ I_x(u,x,A) \, 2 \int_0^1 J(0,x,u') \, du' \, . \qquad (6.64)$$

Next we shall consider the function $I$ defined in Eq. (6.58). Through differentiation we find that

$$I_x(u,x,A) = A \left\{ - e^{-x/u} + 2J(x,x,u,A) \, E_2(x) \right.$$

$$+ 2 \int_0^x \left[ - \frac{1}{u} J(y,x,u,A) \right.$$

$$+ 2J(x,x,u,A) \int_0^1 J(y,x,u',A) \, du'/u' \Big] E_2(y) \, dy \Big\}$$

or

$$I_x(u,x,A) = - \frac{1}{u} I(u,x,A) + 2J(x,x,u,A) \int_0^1 I(u',x,A) \, du'/u' \, .$$
$$(6.65)$$

This is the desired differential equation for the function $I$.

The differential equation for the function $J(0,x,u,A)$ follows from Eq. (6.55) by putting $t = 0$. It is

$$J_x(0,x,u,A) = -\frac{1}{u} J(0,x,u,A)$$

$$+ 2J(x,x,u,A) \int_0^1 J(0,x,u',A) \, du'/u' . \qquad (6.66)$$

Lastly, we note that the functions $J(0,x,u)$ and $J(x,x,u)$ satisfy the differential equations

$$\frac{d}{dx} J(x,x,u) = J(0,x,u) \, 2 \int_0^1 J(0,x,u') \, du'/u' \qquad (6.67)$$

and

$$\frac{d}{dx} J(0,x,u) = -\frac{1}{u} J(0,x,u)$$

$$+ 2J(x,x,u) \int_0^1 J(0,x,u') \, du'/u' . \qquad (6.68)$$

These equations follow from Eqs. (6.64) and (6.66) by putting $A = 0$ and from the fact that

$$I(u,x,0) = 0 , \qquad (6.69)$$

which is evident from Eq. (6.58).

The needed differential equations for the functions $J(t,x,u,A)$ and the auxiliary functions $J(0,x,u,A)$, $J(x,x,u,A)$, $J(0,x,u)$, $J(x,x,u)$ and $I(u,x,A)$ have now been obtained. They are given in Eqs. (6.55), (6.64),

(6.65) - (6.68). The initial conditions at $x = 0$ are

$$J(0,0,u,A) = \frac{1}{4} + \frac{1}{2} Au , \qquad (6.70)$$

$$J(0,0,u) = \frac{1}{4} , \qquad (6.71)$$

and

$$I(u,0,A) = Au . \qquad (6.72)$$

These follow from Eqs. (6.51), (6.63), and (6.58).

We introduce the standard nomenclature

$$X(x,u) = 4J(x,x,u) , \qquad (6.73)$$

$$Y(x,u) = 4J(0,x,u) , \qquad (6.74)$$

as well as the new functions

$$X(x,u,A) = J(x,x,u,A) , \qquad (6.75)$$

$$Y(x,u,A) = J(0,x,u,A) . \qquad (6.76)$$

The Cauchy system for these four functions and the fifth function $I$ is

$$X_x(x,u) = \frac{1}{2} Y(x,u) \int_0^1 Y(x,u') \, du'/u' , \qquad (6.77)$$

$$Y_x(x,u) = -\frac{1}{u} Y(x,u) + \frac{1}{2} X(x,u) \int_0^1 Y(x,u') \, du'/u' \, , \tag{6.78}$$

$$X_x(x,u,A) = \frac{1}{2} \left[ Y(x,u,A) - I(u,x,A) \right] \int_0^1 Y(x,u') \, du'/u'$$

$$+ \frac{1}{2} \left[ -\frac{1}{u} I(u,x,A) \right.$$

$$+ 2X(x,u,A) \int_0^1 I(u',x,A) \, du'/u' \left. \right] \int_0^1 Y(x,u') \, du' \, , \tag{6.79}$$

$$Y_x(x,u,A) = -\frac{1}{u} Y(x,u,A)$$

$$+ 2X(x,u,A) \int_0^1 Y(x,u',A) \, du'/u' \, , \tag{6.80}$$

and

$$I_x(u,x,A) = -\frac{1}{u} I(u,x,A)$$

$$+ 2X(x,u,A) \int_0^1 I(u',x,A) \, du'/u' \, , \tag{6.81}$$

$$x \geq 0 \, ;$$

$$X(0,u) = 1 \, , \tag{6.82}$$

$$Y(0,u) = 1 \, , \tag{6.83}$$

$$X(0,u,A) = \frac{1}{4} + \frac{1}{2} Au \, , \tag{6.84}$$

$$Y(0,u,A) = \frac{1}{4} + \frac{1}{2} Au \, , \tag{6.85}$$

$$I(u,0,A) = Au .$$                                      (6.86)

The Cauchy system for the function $J(t,x,u,A)$, valid for $x \geq t$, is

$$J_x(t,x,u,A) = -\frac{1}{u} J(t,x,u,A)$$

$$+ 2X(x,u,A) \int_0^1 J(t,x,u',A) \, du'/u' ,$$

$$x \geq t , \quad (6.87)$$

the initial condition at $x = t$ being

$$J(t,t,u,A) = X(t,u,A) .$$                              (6.88)

The Cauchy system lends itself well to numerical solution via the method of lines. The integrals are approximated via finite sums using Gaussian quadrature formulas. Thus the differential-integral equations are reduced to ordinary differential equations with known initial conditions. The reduction of variables is important computationally and leads to interesting, theoretical questions.

## 6.3   SPECULAR REFLECTORS

The purpose of this section is to study the multiple scattering of radiation in an atmosphere which absorbs and isotropically scatters incident radiation, the bottom surface of the atmosphere being a specular reflector. These considerations will help in the interpretation of satellite measurements made in the ultraviolet region when the

air-sea interface may be regarded as a specular reflector.
This applies to ozone determinations from satellite observations and in the estimation of cloud heights from diffusely reflected light measurements.

Consider a plane-parallel isotropically scattering
atmosphere of optical thickness  x  whose lower boundary
is a specular reflector of reflectivity  $\rho(u)$.  The function  $\rho(u)$  denotes the probability that a particle
striking the bottom going in the direction  arccos   u
will be specularly reflected rather than absorbed.  Let
$\lambda$  represent the albedo for single scattering and assume
that the medium is illuminated at the top by parallel rays
of net flux  $\pi$  in the direction arccos  u .  Define the
source function  $J(t,x,u)$  to be the total rate of production of particles per unit volume per unit solid angle at
the altitude  t  above the bottom surface,  $0 \leq t \leq x$ .

The source function by  J  is a solution of the
Fredholm integral equation

$$J(t,x,u) = \frac{\lambda}{4} e^{-(x-t)/u} + \rho(u) e^{-(x+t)/u}$$

$$+ \frac{\lambda}{2} \int_0^x [E_1(|t-y|) + f(t+y)] \, J(y,x,u) \, dy \, ,$$
$$0 \leq t \leq x \, , \quad (6.89)$$

where

$$E_1(s) = \int_0^1 e^{-s/z} z^{-1} \, dz \, , \quad\quad\quad\quad (6.90)$$

$$f(s) = \int_0^1 e^{-s/z} \rho(z) z^{-1} \, dz \, , \quad\quad s>0 \, . \quad (6.91)$$

Let

$$r(v,u,x) = \text{the intensity of the diffusely reflected}$$
$$\text{radiation with angle of reflection arccos}$$
$$v \, . \qquad (6.92)$$

The reflection function $r$ is expressed in terms of the source function $J$ by means of the relation

$$r(v,u,x) = v^{-1} \int_0^x J(y,x,u) \left[ e^{-(x/y)/v} + \rho(v)e^{-(x+y)/v} \right] dy.$$
$$(6.93)$$

Let

$$R(v,u,x) = \int_0^x J(y,x,u) \left[ e^{-(x-y)/v} + \rho(v)\, e^{-(x+y)/v} \right] dy \, ,$$
$$(6.94)$$

so that

$$R(v,u,x) = vr(v,u,x) \, . \qquad (6.95)$$

The transmission function $t(v,u,x)$ is considered to be

$$t(v,u,x) = \text{the intensity of the diffuse radiation at}$$
$$\text{the bottom of the atmosphere in a direc-}$$
$$\text{tion making an angle arccos } v \text{ with}$$
$$\text{respect to a downward normal to the specular}$$
$$\text{reflector.} \qquad (6.96)$$

The transmission function $t(v,u,x)$ is not to be confused with optical height $t$.

In terms of the source function J, the function t is expressed as

$$t(v,u,x) = v^{-1} \int_0^x e^{-y/v} J(y,x,u) \, dy \ . \qquad (6.97)$$

We also introduce the function T to be

$$T(v,u,x) = \int_0^x e^{-y/v} J(y,x,u) \, dy \ , \qquad (6.98)$$

so that

$$T(v,u,x) = vt(v,u,x) \qquad (6.99)$$

We shall now derive initial value problems for the functions R and T, x being viewed as the independent variable.
Differentiate Eq. (6.94) with respect to x,

$$R_x(v,u,x) = J(x,x,u) \left[ 1 + \rho(v) \ e^{-2x/v} \right] - \frac{1}{v} R(v,u,x)$$

$$+ \int_0^x J_x(y,x,u) \left[ e^{-(x-y)/v} + \rho(v) \ e^{-(x+y)/v} \right] dy. \qquad (6.100)$$

To obtain an expression for $J_x(y,x,u)$, differentiate Eq. (6.89) with respect to x to obtain

$$J_x(t,x,u) = -\frac{1}{u} \frac{\lambda}{4} \left[ e^{-(x-t)/v} + \rho(u) \ e^{-(x+t)/u} \right]$$

$$+ \frac{\lambda}{2} \left[ E_1(x-t) + f(t+x) \right] J(x,x,u) \qquad (6.101)$$

$$+ \frac{\lambda}{2} \int_0^x \left[ E_1(|t-y|) + f(t+y) \right] J_x(y,x,u) \, dy.$$

Regard Eq. (6.101) as an integral equation for the function $J_x(t,x,u)$ with two forcing terms. It follows from Eqs. (6.89) - (6.91) that the solution $J_x$ may be expressed in the form

$$J_x(t,x,u) = -\frac{1}{u} J(t,x,u)$$

(6.102)

$$+ 2J(x,x,u) \int_0^1 J(t,x,z) \, dz/z \, , \quad x>t \, ,$$

which is also a differential equation for $J$. Equation (6.100) for $R$ becomes

$$R_x(v,u,x) = J(x,x,u) \left[ 1 + \rho(v) \, e^{-2x/v} \right] - \frac{1}{v} R(v,u,x)$$

$$+ \int_0^x \left[ -\frac{1}{u} J(y,x,u) + J(x,x,u) \, \Phi(y,x) \right]$$

$$\cdot \left[ e^{-(x-y)/v} + \rho(v) \, e^{-(x+y)/v} \right] dy \, , \quad (6.103)$$

where

$$\Phi(t,x) = 2 \int_0^1 J(t,x,u) \, du/u \, . \tag{6.104}$$

This equation reduces to

$$R_x(v,u,x) = -\left( \frac{1}{u} + \frac{1}{v} \right) R(v,u,x) + J(x,x,u)$$

$$\cdot \left[ 1 + \rho(v) \, e^{-2x/v} \right.$$

$$+ \int_0^x \Phi(y,x) \left\{ e^{-(x-y)/v} + \rho(v)^{-(x+y)/v} \right\} dy \right].$$

(6.105)

Making use of Eqs. (6.104) and (6.94) it is seen that

$$R_x(v,u,x) = - \left(\frac{1}{u} + \frac{1}{v}\right) R(v,u,x) + J(x,x,u)$$

$$\cdot \left[1 + \rho(v) \; e^{-2x/v} + 2 \int_0^1 R(v,u',x) \; du'/u'\right].$$

(6.106)

To express $J(x,x,u)$ in terms of $R$ we return to Eq. (6.89) and replace $t$ by $x$,

$$J(x,x,u) = \frac{\lambda}{4} \left[1 + \rho(u) \; e^{-2x/u}\right]$$

$$+ \frac{\lambda}{2} \int_0^x \left[E_1(x-y) + f(x+y)\right] J(y,x,u) \; dy .$$

(6.107)

Keeping in mind Eqs. (6.90), (6.91), and (6.94), we see that this becomes

$$J(x,x,u) = \frac{\lambda}{4} \left[1 + \rho(u) \; e^{-2x/u}\right] + \frac{\lambda}{2} \int_0^1 R(v',u,x) \; dv'/v'$$

$$= \frac{\lambda}{4} \left[1 + \rho(u) \; e^{-2x/u} + 2 \int_0^1 R(v',u,x) \; dv'/v'\right].$$

(6.108)

The final form of the differential equation for $R$ is

$$R_x(v,u,x) = -\left(\frac{1}{u} + \frac{1}{v}\right) R(v,u,x)$$

$$+ \frac{\lambda}{4} \left[1 + \rho(u) \, e^{-2x/u} + 2 \int_0^1 R(v',u,x)dv'/v'\right]$$

$$\cdot \left[1 + \rho(v) \, e^{-2x/v} + 2 \int_0^1 R(v,u',x)du'/u'\right],$$

$$\tag{6.109}$$

$$x \geq 0 \ .$$

The initial condition at $x = 0$, according to the definition
of the function $R$ in Eq. (6.94) is

$$R(v,u,0) = 0 \ . \tag{6.110}$$

The Cauchy problem in Eqs. (6.109) and (6.110) determines the
function $R$ for $0 \leq x$ and $0 \leq u,v \leq 1$ .

The Cauchy problem for the source function is given
by differential Eq. (6.102), where $J(x,x,u)$ is expressed
as in Eq. (6.108), and the initial condition at $x = t$,

$$J(t,t,u) = \frac{\lambda}{4} \left[1 + \rho(u) \, e^{-2t/u} + 2 \int_0^1 R(v',u,t)dv'/v'\right].$$

$$\tag{6.111}$$

In these equations, the function $R$ may be regarded as
known through its Cauchy problem.

Now differentiate Eq. (6.98) with respect to $x$ to
obtain

$$T_x(v,u,x) = e^{-x/v} J(x,x,u) + \int_0^x e^{-y/v} J_x(y,x,u) \, dy \ .$$

$$\tag{6.112}$$

According to Eq. (6.102) this becomes

$$T_x(v,u,x) = e^{-x/v} J(x,x,u)$$

$$+ \int_0^x e^{-y/v} \left[ -\frac{1}{u} J(y,x,u) + J(x,x,u)\Phi(y,x) \right] dy,$$

$$(6.113)$$

or

$$T_x(v,u,x) = e^{-x/v} J(x,x,u) - \frac{1}{u} T(v,u,x)$$

$$+ J(x,x,u) \int_0^x e^{-y/v} \Phi(y,x) \, dy \, . \qquad (6.114)$$

Again using the definition of $\Phi(y,x)$ in Eq. (6.104) and the definition of $R$ in Eq. (6.94) it is seen that

$$T_x(v,u,x) = -\frac{1}{u} T(v,u,x)$$

$$+ \frac{\lambda}{4} \left[ 1 + \rho(u) \, e^{-2x/u} + 2 \int_0^1 R(v',u,x) \, \frac{dv'}{v'} \right]$$

$$\cdot \left[ e^{-x/v} + 2 \int_0^1 T(v,u',x) \, \frac{du'}{u'} \right]. \qquad (6.115)$$

Equation (6.98) provides the initial condition at $x = 0$,

$$T(v,u,0) = 0 \, , \qquad\qquad\qquad 0 \le u, v \le 1 \, . \quad (6.116)$$

Statement of the Cauchy Problem. It is convenient to restate the initial value problem that determines the functions $R$, $T$ and $J$ :

$$R_x(v,u,x) = -\left(\frac{1}{u}+\frac{1}{v}\right)R(v,u,x)$$

$$+\frac{\lambda}{4}\left[1+\rho(u)\,e^{-2x/u}+2\int_0^1 R(v',u,x)\frac{dv'}{v'}\right]$$

$$\cdot\left[1+\rho(v)\,e^{-2x/v}+2\int_0^1 R(v,u',x)\frac{du'}{u'}\right],$$

$$x\geq 0\;,\quad (6.117)$$

$$R(v,u,0)=0\;,\qquad\qquad (6.118)$$

$$T_x(v,u,x)=-\frac{1}{u}T(v,u,x)$$

$$+\frac{\lambda}{4}\left[1+\rho(u)\,e^{-2x/u}+2\int_0^1 R(v',u,x)\frac{dv'}{v'}\right]$$

$$\cdot\left[e^{-x/v}+2\int_0^1 T(v,u',x)\frac{du'}{u'}\right],$$

$$x\geq 0\;,\quad (6.119)$$

$$T(v,u,0)=0\;,\qquad\qquad (6.120)$$

$$J_x(t,x,u)=-\frac{1}{u}J(t,x,u)$$

$$+\frac{\lambda}{2}\left[1+\rho(u)\,e^{-2x/u}+2\int_0^1 R(w,u,x)\,dw/w\right]$$

$$\cdot\int_0^1 J(t,x,z)\,dz/z\;,\qquad x\geq t\;,\quad (6.121)$$

$$J(t,t,u)=\frac{\lambda}{4}\left[1+\rho(u)\,e^{-2t/u}\right]+\frac{\lambda}{2}\int_0^1 R(w,u,t)\,dw/w\;.$$

$$(6.122)$$

From Eqs. (6.117) and (6.118) it is clear that R is symmetric in u and v ,

$$R(v,u,x) = R(u,v,x) , \qquad\qquad x \geq 0 . \qquad (6.123)$$

Numerical Method. An effective numerical method for R and T is based on approximating the integrals that occur by means of finite sums using Gaussian quadrature formulas of order N. Let $\alpha_1, \alpha_2, \ldots \alpha_N$ be the appropriate abscissas and $w_1, w_2, \ldots w_N$ the corresponding Christoffel numbers. Let

$$R(\alpha_i, \alpha_j, x) = R_{ij}(x) , \qquad\qquad (6.124)$$

and

$$T(\alpha_i, \alpha_j, x) = T_{ij}(x) , \qquad i, j = 1, 2, \ldots, N. \quad (6.125)$$

The Cauchy problem in Eqs. (6.117) - (6.120) is then approximated by the initial value problem involving $2N^2$ ordinary differential equations

$$R'_{ij} = - \left( \alpha_j^{-1} + \alpha_i^{-1} \right) R_{ij}$$

$$+ \frac{\lambda}{4} \left[ 1 + \rho(\alpha_j) e^{-2x/\alpha_j} + 2 \sum_{m=1}^{N} R_{mj} \alpha_m^{-1} w_m \right]$$

$$\cdot \left[ 1 + \rho(\alpha_i) e^{-2x/\alpha_i} + 2 \sum_{m=1}^{N} R_{im} \alpha_m^{-1} w_m \right],$$

$$x > 0 , \quad (6.126)$$

$$R_{ij}(0) = 0 , \qquad\qquad\qquad (6.127)$$

$$T'_{ij} = - \alpha_j^{-1} T_{ij}$$

$$+ \frac{\lambda}{4} \left[ 1 + \rho(\alpha_j) e^{-2x/\alpha_j} - 2 \sum_{m=1}^{N} R_{mj} \alpha_m^{-1} w_m \right]$$

$$\cdot \left[ e^{-x/\alpha_i} + 2 \sum_{m=1}^{N} T_{im} \alpha_m^{-1} w_m \right] , \quad x \geq 0 , \quad (6.128)$$

$$T_{ij}(0) = 0 , \qquad\qquad i, j = 1,2,\ldots,N . \quad (6.129)$$

To highlight the effects caused by the specular re-flector, numerical experiments are performed for the case

$$\rho(v) \equiv 1.0 , \qquad\qquad 0 \leq v \leq 1 . \quad (6.130)$$

Some of the results obtained are presented in Fig. 6.4 and in Table 6.3. They are obtained using $N = 7$, and an integration of step-size of 0.005.

In Fig. 6.4, we show a typical graph of reflected intensity versus optical thickness. The incident and output angles are arccos $\alpha_7 \cong 13^0$. The albedo for single scattering is $\lambda = 0.1$. The curve of the reflected intensity is plotted for thickness $x$ ranging from zero to 4.0. As expected for this small value of $\lambda$, the curve rises from zero value to a peak, then falls off to a limiting value.

On physical grounds, we expect that this limit is the same as that for the case in which there is no reflector at the bottom. This value is 0.0134, from Ref. 5, p. 54.

Table 6.3 for the specular reflector with $\lambda = 1.0$ and $x = 3.0$ may be compared against Table 6.2 for the same $\lambda$

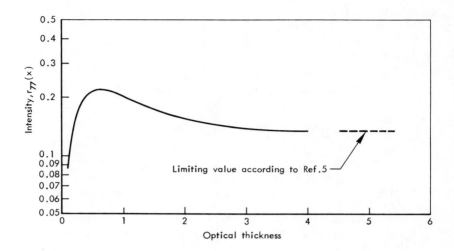

Figure 6.4   Reflected Intensity vs. Optical Thickness
             for  $\lambda = 0.1$  and a Specular Reflector with
             $\rho \equiv 1$

and a Lambert reflector and a larger thickness $(x = 100)$.
Both reflectors have a conservative surface albedo.   Table
6.3 shows values of  $r(v,u,x)$  and  $t(v,u,x)$,  as well as
the total reflected flux, defined through the formula,

Table 6.3

Reflection and Transmission Functions for $\lambda = 1.0$ and $x = 3.0$, for Specular Reflectors with $\rho = 1.0$

Reflected Intensities

| Incident Angle (Deg) | 1(88.5) | 2(82.6) | 3(72.7) | 4(60.0) | 5(45.3) | 6(29.5) | 7(13.0) | Reflected Flux |
|---|---|---|---|---|---|---|---|---|
| 1 | 0.1447 | 0.0579 | 0.0347 | 0.0262 | 0.0224 | 0.0205 | 0.0196 | 0.0799 |
| 2 | 0.2941 | 0.2140 | 0.1623 | 0.1352 | 0.1209 | 0.1132 | 0.1096 | 0.4060 |
| 3 | 0.4056 | 0.3732 | 0.3350 | 0.3070 | 0.2893 | 0.2789 | 0.2736 | 0.9333 |
| 4 | 0.5152 | 0.5232 | 0.5167 | 0.5065 | 0.4977 | 0.4916 | 0.4882 | 1.5708 |
| 5 | 0.6177 | 0.6575 | 0.6845 | 0.6997 | 0.7077 | 0.7113 | 0.7125 | 2.2083 |
| 6 | 0.7001 | 0.7630 | 0.8174 | 0.8562 | 0.8812 | 0.8953 | 0.9016 | 2.7356 |
| 7 | 0.7503 | 0.8266 | 0.8976 | 0.9515 | 0.9878 | 1.0091 | 1.0190 | 3.0616 |

Transmitted Intensities

| Incident Angle ((Deg)) | 1(88.5) | 2(82.6) | 3(72.7) | 4(60.0) | 5(45.3) | 6(29.5) | 7(13.0) | Transmitted Flux |
|---|---|---|---|---|---|---|---|---|
| 1 | 0.0119 | 0.0119 | 0.0119 | 0.0119 | 0.0119 | 0.0118 | 0.0118 | 0.0373 |
| 2 | 0.0734 | 0.0734 | 0.0734 | 0.0736 | 0.0736 | 0.0731 | 0.0725 | 0.2303 |
| 3 | 0.2112 | 0.2113 | 0.2115 | 0.2122 | 0.2118 | 0.2100 | 0.2080 | 0.6626 |
| 4 | 0.4389 | 0.4390 | 0.4397 | 0.4407 | 0.4382 | 0.4324 | 0.4273 | 1.3741 |
| 5 | 0.7295 | 0.7296 | 0.7300 | 0.7287 | 0.7204 | 0.7073 | 0.6968 | 2.2848 |
| 6 | 1.0062 | 1.0061 | 1.0051 | 0.9997 | 0.9838 | 0.9623 | 0.9461 | 3.1673 |
| 7 | 1.1882 | 1.1877 | 1.1855 | 1.1766 | 1.1550 | 1.1276 | 1.1075 | 3.7588 |

$$F_R(u,x) = \pi u \ e^{-2x/u} + 2\pi \int_0^1 r(v,u,x) \ v \ dv \ , \quad (6.131)$$

and the total flux incident on the specular reflector at the bottom,

$$F_T(u,x) = \pi u \ e^{-x/u} + 2\pi \int_0^1 t(v,u,x) \ v \ dv \ . \quad (6.132)$$

The incident angle is constant in each row. Reflection angles are constant in each column. Values of the seven angles are given at the head of each column.

The Numerical Procedure for Source Functions. To deal numerically with the differential-integral equations for $J(t,x,u)$ and $R(v,u,x)$, the integrals appearing are approximated by finite sums. Many numerical quadrature schemes will accomplish this reduction, but past experience indicates that Gaussian quadrature works well.

Denoting $J(t,x,\alpha_j)$ by $J_j(t,x)$, the differential equation (Eq. (6.121)) for $J(t,x,u)$ becomes

$$J_j'(t,x) = -\frac{1}{\alpha_j} J_j(t,x)$$

$$+ \frac{\lambda}{2} \left[ 1 + \rho(\alpha_j) \ e^{-2x/\alpha_j} + 2 \sum_{i=1}^{N} R_{ij}(x) w_i/\alpha_i \right]$$

$$\cdot \left[ \sum_{m=1}^{N} J_m(t,x) w_m/\alpha_m \right],$$

$$j = 1,2,\ldots,N \ , \quad x \geq t \ . \quad (6.133)$$

The functions $R_{ij}(x)$ are solutions of Eqs. (6.126)-(6.127).

Recall in the above expression that the primes indicate differentiation with respect to x, and t is a fixed point
with $0 \leq t \leq x$. The initial conditions for x = t are given
by (see Eq. (6.122))

$$J_j(t,t) = \frac{\lambda}{4} \left[ 1 + \rho(\alpha_j) \, e^{-2t/\alpha_j} \right]$$

$$+ \frac{\lambda}{2} \sum_{i=1}^{N} R_{ij}(t) \, w_i/\alpha_i \, ,$$

$$j = 1,2,\ldots,N \, . \qquad (6.134)$$

To produce the source function at a fixed point t,
the calculation procedure is: Integrate Eq. (6.126) for R
from x = 0 to x = t, using the correct initial conditions.
At this point adjoin Eq. (6.133) with initial conditions
given by Eq. (6.134) to the set for R. Integrate the
augmented set of differential equations from x = t to
$x = x_0$, the desired maximum thickness. The values $J_j(t,x_0)$,
j = 1,2,...,N, will be the desired values of the source
function at the point t.

This procedure is applied to calculate the source
function J. The accuracy of the calculations is checked
by a variety of comparisons, both internal and external.

External checks consist of a qualitative check with
source function curves presented in Chapter 2 for the case
of no reflector at the bottom ($\rho = 0$). A typical graph
for the case $\lambda = 0.9$ is shown in Fig. 6.5. Note that the
curves are qualitatively the same, with the exception of
the small tail near the bottom of the medium. This departure
is to be expected on physical grounds, since no radiation
escapes through the bottom when $\rho = 1$.

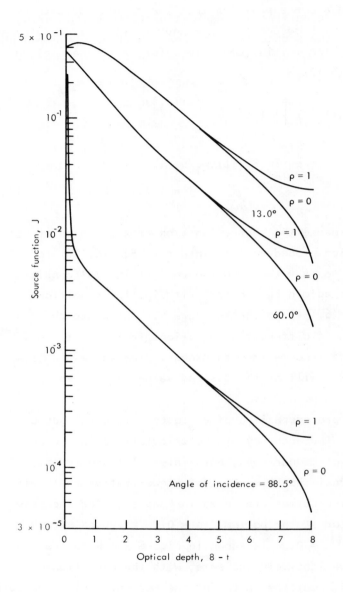

Figure 6.5  Several Source Functions for  x = 8.0
λ = 0.9, and Specular Reflectors with
ρ = 0   and   ρ = 1

A semi-quantitative check is provided by comparing the values of the source function near the top of the medium for $\rho = 1.0$ with those for $\rho = 0.0$. For a reasonably thick medium and $\lambda < 1$, the presence of the reflector at the bottom will have little effect on the source functions near the top of the medium, since little radiation will have been able to penetrate to the bottom and return to the top to interact. The calculations indeed bear out this conjecture.

The last external check consists of using the relations between the source function $J$ and the reflected and transmitted intensities denoted by $r(v,u,x)$ and $t(v,u,x)$, respectively. The relationships are

$$r(v,u,x) = \frac{1}{v} \int_0^x \left[ e^{-(x-y)/v} + \rho(v) \; e^{-(x+y)/v} \right] J(y,x,u) \, dy$$

(6.135)

and

$$t(v,u,x) = \frac{1}{v} \int_0^x e^{-y/v} \; J(y,x,u) \, dy \; .$$

(6.136)

Simpson's Rule with 11 points on the interval $0 \leq y \leq 1$ is applied to evaluate the integrals occurring in Eqs. (6.135) and (6.136) for $u = 0.5$ and all output angles. The source functions used are those generated by the above method for the case $\lambda = 0.1$. The results of these integrations can be compared with the reflected and transmitted intensities presented earlier in this section. Table 6.4 exhibits the comparisons. Notice that excellent agreement is obtained at all angles except the grazing case, angle 1. This is to be expected since the source function drops off rapidly

with depth for emergent angle 1, so that Simpson's rule is
ineffective.

The calculations of Table 6.4 also provide support-
ing evidence for the transmitted intensities presented
earlier.

Table 6.4

Comparison of Reflected and Transmitted Intensities
Calculated Using Simpson's Rule on the Source Function
Against Earlier Results for $\lambda = 0.1$

| Arccos v | $r(v,\frac{1}{2},1)$ | $r(v,1/2,1)$ (Simpson's Rule) | $t(v,\frac{1}{2},1)$ | $t(v,1/2,1)$ (Simpson's Rule) |
|---|---|---|---|---|
| 1 | 0.0252 | 0.0367 | 0.0076 | 0.0107 |
| 2 | 0.0215 | 0.0215 | 0.0081 | 0.0081 |
| 3 | 0.0177 | 0.0177 | 0.0092 | 0.0092 |
| 4 | 0.0153 | 0.0153 | 0.0092 | 0.00092 |
| 5 | 0.0138 | 0.01375 | 0.0086 | 0.0086 |
| 6 | 0.0128 | 0.01275 | 0.0080 | 0.00795 |
| 7 | 0.0122 | 0.0122 | 0.0076 | 0.0076 |

Internal checks consist of varying the step size
of integration and the order of the quadrature N. Calculations have been performed with step size 0.01 and N = 5
and 9, resulting in at least four-digit agreement with the
computations carried out with the standard step size of
0.0005 and quadrature order N = 7.

Discussion. This section has shown that the calculation of the source function for a specular reflector can be
reduced to the solution of a set of ordinary differential
equations with known initial conditions.

The method presented readily extends to the case of
an inhomogeneous atmosphere by letting the parameter $\lambda$ be
a function of altitude. The case of a more general reflecting surface may also be treated by letting the reflectivity
$\rho$ be a function of incident angle rather than a constant.

A fruitful source of future work is the treatment
of inverse problems, wherein the functions $\lambda(t)$ and $\rho(u)$
are to be determined from measurements made on, for example,
the reflected or transmitted intensities. The effects of
polarization and anisotropy should also be included.

## 6.4    EQUIVALENCE RELATIONSHIPS BETWEEN DIFFUSE RADIATION
FIELDS FOR FINITE SLABS BOUNDED BY A PERFECT SPECULAR
REFLECTOR AND A PERFECT ABSORBER

A fundamental goal in scientific investigations is
to reduce a complex problem to one for which solution methods
are known. We will show that certain equivalences exist
between reflected and transmitted intensities and source
functions for a finite slab bounded by a perfect absorber
and for a finite slab bounded by a perfect specular reflector.

Earlier, we have presented numerical results for the
source function and internal and external intensities for the
slab bounded by a perfect absorber, as well as transmitted
and reflected intensities for the slab having a specular
reflector at the bottom. These results will serve as
numerical checks for the simple algebraic formulas to be
presented.

An Equivalence Relation Between Two Source Functions.
Let a homogeneous plane-parallel atmosphere of finite optical
thickness $x > 0$ be illuminated on the top surface by

parallel rays of net flux π incident at an angle arc cos
u with respect to the downward directed normal, $0 \leq u \leq 1$.
Assume the medium absorbs and isotropically scatters radiation, the albedo for single scattering being λ. Consider
the case in which the slab is bounded on the bottom by a
perfect absorber, and denote the source function at optical
altitude t by $J(t,x,u)$. Then J satisfies the integral
equation

$$J(t,x,u) = \frac{\lambda}{4} e^{-(x-t)/u}$$

$$+ \frac{\lambda}{2} \int_0^x E_1(|t-y|) \, J(y,x,u) \, dy \,,$$

$$0 \leq t \leq x \,, \quad (6.137)$$

The function $E_1$ is, as usual, the first exponential integral
function.

Now consider the case in which the medium is bounded
from below by a perfect specular reflector. Throughout this
section, starred quantities refer to the problem with a
specular reflector, unstarred quantities to the problem with
an absorbing boundary. Call $J^*(t,x,u)$ the source function
for this case. $J^*(t,x,u)$ satisfies the integral equation

$$J^*(t,x,u) = \frac{\lambda}{4} \left[ e^{-(x-t)/u} + e^{-(x+t)/u} \right]$$

$$+ \frac{\lambda}{2} \int_0^x \left[ E_1(|t-y|) + E_1(t+y) \right] J^*(y,x,u) dy \,,$$

$$0 \leq t \leq x \,. \quad (6.138)$$

Note that $J^*(t,x,u)$ may be extended to the interval
$-x \leq t \leq 0$ simply by observing that

$$J^*(-t,x,u) = J^*(+t,x,u) \ , \qquad\qquad 0 \leq t \leq x \ , \qquad (6.139)$$

from Eq. (6.138), although no physical interpretation exists
for $J^*(t,x,u)$ for $-x \leq t \leq 0$. However, Eq. (6.139) mathematically allows the extension of the domain of definition
of the function $J^*(t,x,u)$ to the interval $-x \leq t \leq x$.
Making use of the symmetry relation, Eq. (6.139), the
integral equation for $J^*(t,x,u)$ may be written

$$J^*(t,x,u) = \frac{\lambda}{4} \left[ e^{-(x-t)/u} + e^{-(x+t)/u} \right]$$

$$+ \frac{\lambda}{2} \int_{-x}^{x} E_1(|t-y|) \ J^*(y,x,u) \ dy \ ,$$
$$-x \leq t \leq x \ , \qquad (6.140)$$

Comparing Eqs. (6.137) and (6.140), one can see, with a
little algebra, that

$$J^*(t,x,u) = J(x+t,2x,u) + J(x-t,2x,u) \ ,$$
$$-x \leq t \leq x \ , \qquad (6.141)$$

This relationship expresses the source function for the
problem of a perfect specular reflector by means of the
solution to the problem with a perfect absorber.

With regard to the physical model, Eq. (6.141) may
be thought of as arising from a situation in which the
perfect specular reflector is removed, and the medium is
extended from thickness $x$ to thickness $2x$ and illuminated
from the top and bottom by parallel rays of net flux $\pi$
incident at an angle arc cos $u$ with respect to the inward
normal.

Intensity Equivalences. Since the diffuse internal intensities may be expressed in terms of the source function, equivalance formulas may be derived for the intensities, making use of the results for source functions.

When a perfect absorber is at the bottom, the diffuse intensities I(t, +v,x,u) and I(t,-v,x,u) are defined as

$$I(t,+v,x,u) = \frac{1}{v} \int_0^t e^{-(t-y)/v} J(y,x,u) \, dy \qquad (6.142)$$

and

$$I(t,-v,x,u) = \frac{1}{v} \int_t^x e^{-(x-t)/v} J(y,x,u) \, dy \,,$$
$$0 \leq v \leq 1 \,. \qquad (6.143)$$

The + (-) refers to intensity in the upward (downward) direction with direction cosine relative to the upward vertical. Similarly, for the case of the specular reflector, the intensities are expressed in the form

$$I^*(t,+v,x,u) = \frac{1}{v} \int_0^t e^{-(t-y)/v} J^*(y,x,u) \, dy$$

$$+ \frac{1}{v} \int_0^x e^{-(t+y)/v} J^*(y,x,u) \, dy \,, \qquad (6.144)$$

$$I^*(t,-v,x,u) = \frac{1}{v} \int_t^x e^{-(y-t)/v} J^*(y,x,u) \, dy \,,$$
$$0 \leq v \leq 1 \,. \qquad (6.145)$$

Equation (6.141) shows that the intensities in the two cases are related through the equation

$$I^*(t,v,x,u) = I(x+t,v,2x,u) + I(x-t,-v,2x,u) \ ,$$
$$0 \le t \le x; \quad -1 \le v \le 1. \quad (6.146)$$

From Eq. (6.146) the diffusely reflected and transmitted intensities may be expressed by specializing  t
and  v.  Thus,

$$I^*(x,+v,x,u) = R^*(v,u,x), \quad I(2x,v,2x,u) = R(v,u,2x) \ ,$$
$$(6.147)$$

and

$$I^*(0,-v,x,u) = T^*(v,u,x), \quad I(0,-v,2x,u) = T(v,u,2x) \ .$$
$$(6.148)$$

For the diffusely reflected intensity from a slab bounded
by a perfect specular reflector, Eq. (6.146) may be expressed
as

$$R^*(v,u,x) = R(v,u,2x) + T(v,u,2x) \ . \quad\quad (6.149)$$

Now consider that the medium of thickness  x  is
uniformly illuminated at the top by omnidirectional sources
with unit energy per unit horizontal area per unit solid
angle per unit time.  Assume the lower boundary is a perfect
absorber.  The source function is given then by the solution
of the integral equation

$$\Phi(t,x) = \frac{\lambda}{2} E_1(x-t) + \frac{\lambda}{2} \int_0^x E_1(|t-y|)\Phi(y,x) \, dy \, ,$$

$$0 \le t \le x \, . \qquad (6.150)$$

The   b   and   h   functions, defined as

$$b(t,+v,x) = \frac{1}{v} \int_0^t \Phi(y,x) \, e^{-(t-y)/v} \, dy \, , \qquad (6.151)$$

$$b(t,-v,x) = \frac{1}{v} \int_t^x \Phi(y,x) \, e^{-(y-t)/v} \, dy + \frac{1}{v} \, e^{-(x-t)/v}$$

$$(6.152)$$

$$h(t,\pm v,x) = b(x-t,\mp v,x) \, , \qquad\qquad 0 \le v \le 1 \, , \qquad (6.153)$$

are the internal intensities for this case. The  h  functions correspond to intensities when the omnidirectional
sources are at the bottom rather than at the top.

Chapter 3 shows that the  J  and  I  functions (for
the case of parallel rays and an absorbing boundary) may be
expressed algebraically in terms of the  b  and  h  functions.
From these and Eq. (6.146) we obtain

$$J^*(t,x,u) = \frac{\lambda u}{4} \, [1 + ub(2x,u,2x)] \, [b(x+t,-u,2x)$$

$$+ \, b(x-t,-u,2x)]$$

$$+ \frac{\lambda u}{4} \, [uh(2x,u,2x)] \, [h(x+t,-u,2x)$$

$$+ \, h(x-t,-u,2x)] \, . \qquad (6.154)$$

To deal with the internal intensities $I^*(t, \pm v, x, u)$, again
a formula from Chapter 3 is used, as well as Eq. (6.146)
yielding

$$
I^*(t,v,u,x) = \frac{\lambda}{4}[1+ub(2x,u,2x)]\left\{\left(\frac{1}{u}+\frac{1}{v}\right)^{-1}\frac{u}{v}b(x+t,-u,2x)\right.
$$

$$
+b(x+t,v,2x) +\left(\frac{1}{u}-\frac{1}{v}\right)^{-1}\left[-\frac{u}{v}b(x-t,-u,2x)+b(x-t,-v,2x)\right]\Big\}
$$

$$
- \frac{\lambda}{4}uh(2x,u,2x)\left\{\left(\frac{1}{u}+\frac{1}{v}\right)^{-1}\left[\frac{u}{v}h(x+t,-u,2x)\times h(x+t,v,2x)\right]\right.
$$

$$
+ \left(\frac{1}{u}-\frac{1}{v}\right)^{-1}\left[-\frac{u}{v}h(x-t,-u,2x) + h(x-t,-v,2x)\right]\Big\}
$$

$$
-1\le v\le 1 . \quad (6.155)
$$

Equations (6.154) and (6.155) show that the solution
to the problem of monodirectional illumination of a slab
bounded by a perfect specular reflector can be given solely
in terms of the solution to the problem of omnidirectional
illumination of a slab bounded by a perfect absorber.

Numerical Verifications. Reflected and transmitted
intensities and fluxes for a slab bounded by a specular
reflector were presented. Equations (6.147)-(6.149) and
(6.155) provide a check of the earlier results. For example,
when $\lambda = 1.0$, $x = 0.5$, $u = 0.5$ and $v = 0.9746$, the diffusely transmitted intensity computed by Eq. (6.155) is
0.2231, which is in four-figure agreement with the earlier
result. Other comparisons are shown in Tables 6.5 and 6.6.

Discussion. This section has shown that, for some
scattering processes, reducing the solution of a complex
process to the solution of a simpler model is possible.
This reduction has obvious analytic and computational

Table 6.5

Comparison of Two Calculations for Intensities Diffusely
Reflected from a Slab with a Perfect Specular Reflector
at the Bottom for  $\lambda = 1.0$ , x = 0.5

| v | u | $R^*(v,u,x)$ | $R(v,u,2x) + T(v,u,2x)$ (Ref. 3) |
|---|---|---|---|
| 0.026 | 0.026 | 0.1448 | 0.1449 |
| 0.026 | 0.974 | 0.0184 | 0.0184 |
| 0.500 | 0.026 | 0.5323 | 0.5323 |
| 0.500 | 0.974 | 0.3560 | 0.3560 |
| 0.974 | 0.026 | 0.7050 | 0.7051 |
| 0.974 | 0.974 | 0.5179 | 0.5179 |

Table 6.6

Comparison of Two Calculations for Intensities Diffusely
Reflected from a Slab with a Perfect Specular Reflector
at the Bottom for  $\lambda = 0.1$ , x = 0.5

| v | u | $R^*(v,u,x)$ | $R(v,u,2x) + T(v,u,2x)$ (Ref. 3) |
|---|---|---|---|
| 0.026 | 0.026 | 0.0126 | 0.0126 |
| 0.500 | 0.500 | 0.0202 | 0.0202 |
| 0.974 | 0.974 | 0.0214 | 0.0214 |

advantages.  For example, in the calculations performed
earlier for a perfect specular reflector, the reflected and
transmitted intensities reached saturation levels at much
smaller thicknesses than intuition and previous experience
could account for.  However, the insight gained from the
equations now offers the explanation that the process of
inserting a perfect specular reflector at the boundary has
essentially the same effect as doubling the thickness of
the medium.  Hopefully, formulas analogous to those presented
here will be developed for other situations and will give
similar insight into the structure of the scattering process.

Even though the results are highly specialized from
a radiative transfer viewpoint, since they concern only the
case of a perfect reflector, they are important in the general
treatment of Fredholm integral equations having composite
kernels of the form  $k(t,y) = k_1(|t-y|) + k_1(|t+y|)$.  The
equivalence relations presented here provide a means of
reducing the computational burden of the initial value
method presented in Ref. 1 for treating such integral equations.

Finally, the kinship between these results and the
earlier reduction of spherical problems to planar problems
by Heaslet and Warming, Ref. 6., should be noted.


## 6.5    INVERSE PROBLEMS


Attention is now focused on inverse problems in
which the properties of atmospheres and surfaces based on
radiation measurements are sought. Parallel rays of light
of net flux  $\pi$  are incident on a plane-parallel inhomogeneous atmosphere, at the bottom of which is found a

surface reflecting radiation according to Lambert's law.
Given measurements of the angular distribution of the diffusely reflected radiation, it is required to estimate the
optical thickness $x_0$ of the atmosphere, the albedo $\lambda$ as
a function of height, and the albedo $A$ of the Lambert
surface. Such an inverse problem can be routinely treated
using the quasilinearization technique.

Specifically, the case is considered in which $\lambda$ has
the form

$$\lambda(t) = 0.5 + at + bt^2 , \qquad\qquad 0 \leq t \leq x_0 , \qquad (6.156)$$

where $a$ and $b$ are constants. Let the (approximate)
measurements be $\beta_{ij} \approx r_{ij}(x_0) = r(\alpha_i, \alpha_j, x_0)$, where
$\alpha_1, \alpha_2, \ldots, \alpha_N$ are abscissas for Gaussian quadrature with
$N = 7$ points. Values of the parameters $x_0$, $a, b$ and $A$
are required such that the solution of the equations

$$R'_{ij} = - (\alpha_i^{-1} + \alpha_j^{-1}) R_{ij}$$

$$+ \lambda(x) \left[ 1 + 0.5 \sum_{k=1}^{N} R_{ik} w_k/\alpha_k \right]$$

$$\cdot \left[ 1 + 0.5 \sum_{k=1}^{N} R_{kj} w_k/\alpha_k \right] , \qquad (6.157)$$

with initial values

$$R_{ij}(v) = 4A\alpha_i\alpha_j , \qquad\qquad (6.158)$$

where $\lambda(x)$ is given by Eq. (6.156), gives a minimum of
the sum of squares of deviations,

$$\sum_{ij} \left\{ R_{ij}(x_0) - 4\alpha_i \beta_{ij} \right\}^2 \tag{6.159}$$

This is a nonlinear boundary value problem which is solved
computationally using quasilinearization. Initial estimates
of the parameters are sequentially refined, the sequence
usually being rapidly convergent.

In the first numerical experiment, 49 accurate
measurements are given for the case in which the true values
of the parameters are $x_0 = 1.0$, a = 2.0, b = - 2.0, and
A = 0.5. The parameters are initially estimated to be
$x_0 = 1.5$, a = 2.2, b = - 1.8, and A = 0.2; these are then
refined. The results of a two-minute calculation on an
IBM 7044 are given in Table 6.7.

Table 6.7

Sequence of Parameter Estimates in a Controlled Experiment

| Iteration | $b_0$ | a | b | A |
|---|---|---|---|---|
| 0 | 1.5 | 2.2 | -1.8 | 0.2 |
| 1 | 1.28 | 1.29 | -1.05 | 0.73 |
| 2 | 1.03 | 1.82 | -1.65 | 0.45 |
| 3 | 0.999 | 1.995 | -1.98 | 0.501 |
| True Values | 1.0 | 2.0 | -2.0 | 0.5 |

Next, a series of ten experiments is performed, involving 49 noisy measurements per experiment. The relative amount of noise in each measurement is a random number drawn from a Gaussian distribution with zero mean and 0.01 standard deviation. The estimates of each trial vary from good (less than 1 percent error) to bad (about 30 percent error). However, the estimates of $x_0$, a, b, and A averaged over the ten trials are in error by only 1 to 7 peccent.

This study has dealt with only one of the many interesting questions regarding the determination of planetary atmospheres and surfaces from various measurements obtained from earth stations or planetary probes. The value of inversion procedures such as this is obvious for the analysis of the experimental data. In addition, computational investigations during the planning of the experiments serve to indicate the type and quality of measurements required for satisfactory estimations.

## EXERCISES

1.  Physically derive the initial value problems for the
    functions  R,  T  and  J  for the case of a Lambert's
    law reflector.

2.  Write a computer program to calculate the reflected and
    transmitted intensities and fluxes for the case of a
    Lambert's law reflector.  Reproduce the results presented
    in this chapter.

3.  Write a program for the source function for a Lambert's
    law reflector.

4.  Repeat Exercise 1 for the case of a specular reflector.

5.  Repeat Exercise 2 for the case of a specular reflector.

6.  Repeat Exercise 3 for the case of a specular reflector.

7.  Design and perform experiments for estimating properties
    of reflecting surfaces based on measurements of dif-
    fusely reflected radiation.

## REFERENCES

1.  Kagiwada, H. and R. Kalaba, *Integral Equations via
    Imbedding Methods*, Addison-Wesley Publishing Co.,
    Reading, Mass., 1974.

2.  Kahle, A. B., "Global Radiation Emerging from a
    Rayleigh-Scattering Atmosphere of Large Optical Thick-
    ness," *Astrophys. J.*, Vol. 151, February 1968, pp.
    637-645.

3.  Kagiwada, H. H., R. E. Kalaba, and R. L. Segerblom,
    "Flux Equivalence Among Rayleigh, Isotropic and Other
    Scattering Models," *J. of Computational Physics*,
    Vol. 3, No. 2, October 1968, pp. 159-166.

4.  Casti, J., H. Kagiwada, and R. Kalaba, "External
    Radiation Fields for Isotropically Scattering Finite
    Atmospheres Bounded by a Lambert Law Reflector," *J.
    Quant. Spectrosc. Radiat. Transfer*, v. 10, 1970, pp.
    637-651.

5.  Bellman, R., R. Kalaba, and M. Prestrud, "Invariant
    Imbedding and Radiative Transfer in Slabs of Finite
    Thickness, *American Elsevier*, New York, 1963.

6.  Heaslet, M. and R. Warming, *J. Quant. Spectrosc. Radiat.
    Transfer*, v. 5, 1969, 669.

CHAPTER 7

ANISOTROPIC SCATTERING

7.1    THE BASIC INTEGRAL EQUATION AND CAUCHY SYSTEM

It is well known that radiative transfer processes
in the haze and clouds of the earth's atmosphere, as well
as in the sea, are highly anisotropic.  The treatment is
extended in this section to radiative transfer in inhomogeneous, anisotropically scattering slabs.  Assume for
simplicity that the lower boundary is a perfect absorber.
As usual, let  x  be the optical thickness,  t  the
optical height,  $\lambda(t)$  the albedo for single scattering,
u  the direction cosine of incident monodirectional illumination, the constant net flux being  $\pi$  per unit normal
area,  v  the cosine of the polar angle of the direction of
propagation at a general point.  In general, there will be
a dependence of the radiation field on  $\phi$ , the azimuth
angle  $(0 \leq \phi \leq 2\pi)$.  We define the azimuth of the incident
radiation to be zero , or  $\phi_o = 0$ .

The phase function  $p(t,v',\phi',v,\phi)$  describes the
consequence of a single scattering event.  Recall that  $\lambda$

is the probability that a photon, which interacts with a
scatterer, is reemitted at all. Suppose that the direction
of the incident photon is given by the angular parameters
$(v,\phi)$, and that for the scattered photon it is $(v',\phi')$. Then

> $p(t,v',\phi',v,\phi)\ dv'\ d\phi'$  is the probability
> that the scattered photon will go into the
> direction interval given by $(v',\phi')$ and
> $(v' + dv',\ \phi' + d\phi')$, at height  t.                (7.1)

Since the probability of the direction of the scattered
photon being somewhere in the total solid angle must be
unity, the normalization condition is

$$\frac{1}{4\pi} \int_{-1}^{1} \int_{0}^{2\pi} p(t,v',\phi',v,\phi)\ dv'd\phi' = 1 \ . \qquad (7.2)$$

The local reciprocity principle is assumed,

$$p(t,v,\phi,v',\phi') = p(t,v',\phi',v,\phi) \ , \qquad (7.3)$$

$$p(t,v',\phi',-v,\phi) = p(t,-v',\phi',v,\phi) \ , \qquad (7.4)$$

the first equation indicating an invariance under a
reversal of paths, and the second under an interchange of
the upward and downward directions.

The source function  $J(t,v,\phi,u)$  is defined so that

> $J(t,v,\phi,u)\ dv\ d\phi$ = the rate of production of
> scattered photons per unit volume at altitude
> t, in the direction interval $(v,\phi)$ to

$(v + dv, \phi + d\phi)$ , the incident direction
having polar angle arccos $u$ .                                    (7.5)

The source function satisfies the formula, for a small
volume located between $t$ and $t + \Delta$ ,

$$J(t,v,\phi,u) \, dvd\phi\Delta = \pi u \cdot e^{-(x-t)/u}$$

$$\cdot \frac{\Delta}{u} \cdot \lambda(t)p(t,v,\phi,-u,\phi_0)dvd\phi \frac{1}{4\pi}$$

$$+ \int_0^1 \int_0^{2\pi} \int_t^x J(y,-v',\phi',u,\phi_0)dv'd\phi' \, dy$$

$$\cdot e^{-(y-t)/v'} \cdot \frac{\Delta}{v'} \cdot \lambda(t) \, p(t,v,\phi,-v'\phi')dvd\phi \frac{1}{4\pi}$$

$$+ \int_0^1 \int_0^{2\pi} \int_0^t J(y,+v',\phi',u,\phi_0) \, dv'd\phi'dy$$

$$\cdot e^{-(t-y)/v'} \cdot \frac{\Delta}{v'} \cdot \lambda(t)p(t,v,\phi,+v',\phi')dvd\phi \frac{1}{4\pi}$$

$$+ o(\Delta) \, .                                                    (7.6)$$

As we let $\Delta$ tend to 0, we obtain the Fredholm integral
equation,

$$J(t,v,\phi,u) = \frac{\lambda(t)}{4} \, p(t,v,\phi,-u,0) \, e^{-(x-t)/u}$$

$$+ \frac{\lambda(t)}{4\pi} \int_0^1 \int_0^{2\pi} \int_t^x J(y,-v',\phi',u) \, e^{-(y-t)/v'}$$

$$p(t,v,\phi,-v',\phi') \, \frac{dv'}{v'} \, d\phi' dy$$

$$+ \frac{\lambda(t)}{4\pi} \int_0^1 \int_0^{2\pi} \int_0^t J(y,+v',\phi',u) \, e^{-(t-y)/v'}$$

$$p(t,v,\phi,+v',\phi') \, \frac{dv'}{v'} \, d\phi' dy \; , \qquad (7.7)$$

for $0 \le t \le x$, $-1 \le v \le 1$, $0 \le \phi \le 2\pi$, $0 < u \le 1$ .

We regard the solution as a function of $x$, the thickness, and we denote it $J(t,v,\phi,x,u)$. The integral equation for the source function is

$$J(t,v,\phi,x,u) = \frac{\lambda(t)}{4} \, p(t,v,\phi,-u,0) \, e^{-(x-t)/u}$$

$$+ \frac{\lambda(t)}{4\pi} \int_0^t \int_0^1 \int_0^{2\pi} p(t,v,\phi,+w,\phi') \, J(y,+w,\phi',x,u) e^{-(x-t)/w}$$

$$d\phi' \, \frac{dw}{w} \, dy$$

$$+ \frac{\lambda(t)}{4\pi} \int_t^x \int_0^1 \int_0^{2\pi} p(t,v,\phi,-w,\phi') \, J(y,-w,\phi',x,u) e^{-(y-t)/w}$$

$$d\phi' \, \frac{dw}{w} \, dy \; , \qquad (7.8)$$

for $-1 \le v \le 1$, $0 \le \phi \le 2\pi$, $0 < u \le 1$, $0 \le t \le x \le x_1$ .

To obtain a Cauchy problem for $J$, we begin by differentiating throughout Eq. (7.8) with respect to $x$. This yields

$$J_x(t,v,\phi,x,u) = -u^{-1} \frac{\lambda(t)}{4} p(t,v,\phi,-u,0)e^{-(x-t)/u}$$

$$+ \frac{\lambda(t)}{4\pi} \int_0^t \int_0^1 \int_0^{2\pi} p(t,v,\phi,+w,\phi') \, J_x(y,+w,\phi',x,u)$$

$$e^{-(t-y)/w} \, d\phi' \, \frac{dw}{w} \, dy$$

$$+ \frac{\lambda(t)}{4\pi} \int_t^x \int_0^1 \int_0^{2\pi} p(t,v,\phi,-w,\phi') \, J_x(y,-w,\phi',x,u)$$

$$e^{-(y-t)/w} \, d\phi' \, \frac{dw}{w} \, dy$$

$$+ \frac{\lambda(t)}{4\pi} \int_0^1 \int_0^{2\pi} p(t,v,-w,\phi') \, J(x,-w,\phi',x,u)$$

$$e^{-(x-t)/w} \, d\phi' \, \frac{dw}{w} \, . \tag{7.9}$$

Regard Eq. (7.9) as an integral equation for the function $J_x(t,v,\phi,x,u)$. Its solution is

$$J_x(t,v,\phi,x,u) = -u^{-1} J(t,v,\phi,x,u)$$

$$+ \pi^{-1} \int_0^1 \int_0^{2\pi} J(x,-w,\phi',x,u) \, J(t,v,\phi-\phi',w,x)$$

$$d\phi' \, \frac{dw}{w} \, , \tag{7.10}$$

a differential-integral equation for the function $J(t,v,\phi,x,u)$.

Note that the source function at the top appears in the above equation. By setting $t = x$ in the integral equation for $J$, we obtain the formula

$$J(x,v,\phi,x,u) = \frac{\lambda(x)}{4} p(x,v,\phi,-u,0)$$

$$+ \frac{\lambda(x)}{4\pi} \int_0^x \int_0^1 \int_0^{2\pi} p(x,v,\phi,w,\phi')$$

$$J(y,w,\phi',x,u)e^{-(x-y)/w} d\phi' \frac{dw}{w} dy \ . \tag{7.11}$$

Introduce the reflected intensity function $r(v,\phi,u,x)$ and reflection function $R(v,\phi,u,x)$ through the equations

$$r(v,\phi,u,x) = v^{-1} \int_0^x J(y,v,\phi,x,u)e^{-(x-y)/v} dy \ , \tag{7.12}$$

$$R(v,\phi,u,x) = 4vr(v,u,x), \qquad\qquad 0<v\leq 1 \ , \tag{7.13}$$

so that

$$R(v,\phi,u,x) = 4 \int_0^x J(y,v,\phi,x,u)e^{-(x-y)/v} dy \ . \tag{7.14}$$

Then the source function at the top is expressed in terms of $R$ as

$$J(x,v,\phi,x,u) = \frac{\lambda(x)}{4} p(v,\phi,-u,0)$$

$$+ \frac{\lambda(x)}{16\pi} \int_0^1 \int_0^{2\pi} p(v,\phi,w,\phi') \ R(w,\phi',u,x) \ d\phi' \frac{dw}{w} \ ,$$

$$-1 \leq v \leq 1 \ . \quad (7.15)$$

Let us next derive an initial value problem for the function R. By differentiation of Eq. (7.14) we obtain

$$R_x(v,\phi,u,x) = -v^{-1} R(v,\phi,u,x) + 4 J(x,v,\phi,x,u)$$

$$+ 4 \int_0^x J_x(y,v,\phi,x,u) \ e^{-(x-y)/v} \ dy \ . \quad (7.16)$$

For $J_x$ we substitute the expression in Eq. (7.10). Then the integral term in Eq. (7.16) becomes

$$-4u^{-1} \int_0^x J(y,v,\phi,x,u) \ e^{-(x-y)/v} dy$$

$$+ \frac{4}{\pi} \int_0^x \int_0^1 \int_0^{2\pi} J(x,-w,\phi',x,u)$$

$$J(y,v,\phi-\phi',w,x) \ d\phi' \ \frac{dw}{w} e^{-(x-y)/v} \ dy$$

$$= - u^{-1} R(v,\phi,u,x)$$

$$+ \frac{1}{\pi} \int_0^1 \int_0^{2\pi} J(x,-w,\phi',x,u) \ R(v,\phi-\phi',w,x) d\phi' \ \frac{dw}{w} \ . $$

$$(7.17)$$

Using the above expression and the formula for $J(x,v,\phi,x,u)$ , we arrive at the final form of Eq. (7.16),

$$R_x(v,\phi,u,x) = -(v^{-1}+u^{-1}) \, R(v,\phi,u,x) + \lambda(x)p(x,v,\phi,-u,0)$$

$$+ \frac{\lambda(x)}{4\pi} \int_0^1 \int_0^{2\pi} p(x,v,\phi,w,\phi') \, R(w,\phi',u,x) \, d\phi' \, \frac{dw}{w}$$

$$+ \frac{\lambda(x)}{4\pi} \int_0^1 \int_0^{2\pi} p(x,-w,\phi',-u,0) \, R(v,\phi-\phi',w,x)d\phi' \frac{dw}{w}$$

$$+ \frac{\lambda(x)}{16\pi^2} \int_0^1 \int_0^{2\pi} \int_0^1 \int_0^{2\pi} p(x,-w,\phi',z,\phi'')$$

$$R(z,\phi'',u,x) \, d\phi'' \, \frac{dz}{z}$$

$$\cdot R(v,\phi-\phi',w,x) \, d\phi' \, \frac{dw}{w} \, ,$$

$$x \geq 0 \, . \qquad (7.18)$$

The initial condition which follows from Eq. (7.14) is

$$R(v,\phi,u,0) = 0 \, . \qquad\qquad\qquad (7.19)$$

Returning to Eq. (7.10), we write it in the form of an integro differential equation,

$$J_x(t,v,\phi,u) \quad = -u^{-1}J(t,v,\phi,x,u)$$

$$+ \frac{\lambda(t)}{4\pi} \int_0^1 \int_0^{2\pi} p(t,-w,\phi',-u,0) \, J(t,v,\phi-\phi',w,x)d\phi' \, \frac{dw}{w}$$

$$+ \frac{\lambda(t)}{16\pi^2} \int_0^1 \int_0^{2\pi} \int_0^1 \int_0^{2\pi} p(t,-w,z,\phi'')R(z,\phi'',u,x) \, d\phi'' \frac{dz}{z}$$

$$\cdot J(t,v,\phi-\phi',w,x) \, d\phi' \, \frac{dw}{w} \, , \qquad\qquad x \geq 0 \, , \qquad (7.20)$$

which follows from use of Eqs. (7.14) and (7.15).

The initial condition is given at  x = t:

$$J(t,v,\phi,t,u) = \frac{\lambda(t)}{4} p(t,v,\phi,-u,0)$$

$$+ \frac{\lambda(t)}{16\pi} \int_0^1 \int_0^{2\pi} p(t,v,\phi,w,\phi')$$

$$R(w,\phi',u,t) \, d\phi' \, \frac{dw}{w} \quad . \tag{7.21}$$

Equations (7.20) and (7.21) determine the source function
at the fixed altitude  t.

These equations for the source and reflection functions in their full generality can be formidable to solve.
Even if  t  takes on a single value, and  x  is the
independent variable, both  J  and  R  are functions of
three angular variables.  An analytical solution would be
beyond hope, and a direct computational attack would tax
even the largest computers available today.  We can ap-
preciate the need for methods of reducing the problem to
more manageable  proportions.

Several methods of reduction are possible.  These
include rational functions approximation of the phase
function which is useful for axially symmetric fields, and
expansion in Legendre polynomials.


7.2     AXIALLY SYMMETRIC RADIATION FIELDS


Rather than treat the most general anisotropic
scattering problem, let us focus our attention on the case
of anisotropic scattering with axially symmetric radiation

fields.  Such a situation arises when uniform conical flux
of radiation is incident on the upper surface of the slab.
Then the radiation field is independent of azimuth.  The
results may be used for omnidirectional illumination, or for
normal, monodirectional illumination.

Let the incident energy be such that  $\pi u$  is the
energy passing through a unit of horizontal area per unit
time.  It is assumed that the albedo for single scattering
and the phase function are both independent of altitude.
Furthermore it is assumed that the phase function depends
only on the cosine of the angle of deviation of the direction of scattering from the direction of incidence, i.e.,
$p = p(\cos \theta)$, where as usual

$$\cos \theta = vv' + (1-v^2)^{1/2} (1-v'^2)^{1/2} \cos \phi , \qquad (7.22)$$

where  $\phi$  is the difference in azimuth.  In view of Eq.
(7.22), we write  $p = p(v',v,\phi)$.  We will find it convenient
to introduce the function  $c(v',v)$  through the formula

$$c(v',v) = \frac{\lambda}{4\pi} \int_0^{2\pi} p(v',v,\phi) \, d\phi . \qquad (7.33)$$

The source function appropriate to this case is
defined such that

> $J(t,v,x,u)dv =$ the rate of production of scattered
> photons per unit volume at altitude  $t$, in the
> direction cosine interval  $(v, v+dv)$  and with all
> azimuths between  $(0,2\pi)$, the incident direction
> having polar angle arccos u  and the incident

radiation being uniform conical flux of energy
$\pi u$ per unit horizontal area, and the slab thickness being x.                                          (7.24)

The integral equation for the axially symmetric
source function may be derived from the description of the
physical situation.  On the other hand, we may start with
the basic integral equation for anisotropic scattering.  We
make the replacements,

$$p(v',\phi',v,\phi) \longrightarrow p(v',v,\phi) \ ,$$                        (7.25)

$$\frac{1}{2\pi} \int_0^{2\pi} J(t,v,\phi,u,\phi_o) \, d\phi \longrightarrow J(t,v,x,u) \ ,$$       (7.26)

which follows from the definitions.  Using also Eq. (7.23),
we obtain the integral equation

$$J(t,v,x,u) = \frac{1}{2} c(v,-u)e^{-(x-t)/u}$$

$$+ \int_t^x \int_0^1 c(v,-v')e^{-(y-t)/v'} \, J(y,-v',x,u)\frac{dv'}{v'} \, dy$$

$$+ \int_0^t \int_0^1 c(v,+v')e^{-(t-y)/v'} \, J(y,+v',x,u)\frac{dv'}{v'} \, dy \ ,$$

$$0 \leq t \leq x, \quad 0 \leq u \leq 1, \quad -1 \leq v \leq +1. \ (7.27)$$

Let us introduce the reflection function,

$r(v,u,x)$ = the intensity of the energy emerging
from the top of the slab with direction cosine v,
                                                       (7.28)

and the symmetric  R  function,

$$r(v,u,x) = \frac{R(v,u,x)}{4v} \; .$$

(7.29)

A relation between  r  and  J  is

$$r(v,u,x) = \frac{1}{v} \int_0^x e^{-(x-y)/v} \, J(y,v,x,u) \; dy \; ,$$

(7.30)

based on their physical meanings.  It is desired to determine
both  J  and  r  for slabs of thickness  x  sufficiently
small that Eq. (7.27) has a unique solution.  Note that the
reflection function for the case of axially symmetric
fields is a function of only three variables, as is the
reflection function for isotropic scattering.

The family of integral equations (7.27) is equivalent
to a Cauchy system which can be readily handled computationally.  First we state the initial value problem, then prove
that the solution of it satisfies the family of integral
equations.  Then numerical aspects are discussed.

Statement of Cauchy System.  Consider the Cauchy
system for the auxiliary function  R ,

$$R(v,u,0) = 0 \; ,$$

(7.31)

$$R_x(v,u,x) = -\left(\frac{1}{u} + \frac{1}{v}\right) R(v,u,x) + 2c(v,-u)$$

$$+ \int_0^1 c(v,v') \ R(v',u,x) \ \frac{dv'}{v'}$$

$$+ 2 \int_0^1 R(v,v',x) \ \frac{dv'}{v'} \left\{\frac{1}{2} c(-v',-u)\right.$$

$$+ \frac{1}{4} \int_0^1 \frac{dv''}{v''} c(-v',v'') \ R(v'',u,x)\bigg\}, \qquad (7\ 33)$$

$$x \geq 0; \quad 0 \leq v,u \leq 1 \ .$$

Also consider the Cauchy system for the function J,

$$J_x(t,v,x,u) = -\frac{1}{u} J(t,v,x,u)$$

$$+ \int_0^1 \left\{c(-v',-u) + \frac{1}{2} \int_0^1 \frac{dv''}{v''} c(-v',v'')\right.$$

$$R(v'',u,x)\bigg\} \ J(t,v,x,v') \ \frac{dv'}{v'} \ ,$$

$$x \geq t; \quad 0 \leq u \leq 1; \quad -i \leq v \leq +1 \ . \quad (7.33)$$

The initial condition on the function J at x = t is

$$J(t,v,t,u) = \frac{1}{2} c(v,-u) + \frac{1}{4} \int_0^1 \frac{dv'}{v'} c(v,v')R(v',u,t),$$

$$0 \leq u \leq 1; \quad -1 \leq v \leq +1 \ . \quad (7.34)$$

It is assumed that this Cauchy system possesses a unique solution, at least for x sufficiently small.

The aim is to show that the Cauchy system above and the integral Eq. (7.27) are equivalent. These equations have been derived previously on physical grounds.

Validation of Cauchy System. Now we shall show that the solution of the Cauchy system for the functions R and J provides a solution of the integral equation. Our first task is to demonstrate that

$$R(v,u,x) = 4\int_0^x e^{-(x-t)/v} J(y,v,x,u)dy ,$$
$$x\geq0; \; 0\leq v,u<1 . \quad (7.35)$$

Let the function Q be defined by the equation

$$Q(v,u,x) = 4\int_0^x e^{-(x-y)/v} J(y,v,x,u) \, dy ,$$
$$x\geq0; \; 0\leq v,u\leq1 . \quad (7.36)$$

It clearly satisfies the initial condition

$$Q(v,u,0) = 0 . \quad (7.37)$$

Furthermore differentiation of both sides of Eq. (7.36) with respect to x shows that

$$Q_x(v,u,x) = -\frac{1}{v} Q(v,u,x) + 4J(x,v,x,u)$$
$$+ 4\int_0^x e^{-(x-y)/v}\left\{-\frac{1}{u} J(y,v,x,u) +\int_0^1 [c(-v',-u)\right.$$
$$+ \frac{1}{2}\int_0^1 \frac{dv''}{v''} c(-v',v'') \; R(v'',u,x)]J(y,v,x,v')$$
$$\left.\frac{dv'}{v'}\right\} \, dy . \quad (7.38)$$

The last equation may be rewritten as

$$Q_x(v,u,x) = -\left(\frac{1}{v} + \frac{1}{u}\right) Q(v,u,x) + 4J(x,v,x,u)$$

$$+ 4\int_0^x c(-v',-u) \frac{dv'}{v'} \int_0^1 e^{-(x-y)/v} J(y,v,x,v')dy$$

$$+ 2\int_0^1\int_0^1 \frac{dv''}{v''}\frac{dv'}{v'} c(-v',v'') R(v'',u,x)$$

$$e^{-(x-y)/v} J(y,v,x,v')dy , \qquad (7.39)$$

$$Q_x(v,u,x) = -\left(\frac{1}{u} + \frac{1}{v}\right) Q(v,u,x)$$

$$+ 2c(v,-u) + \int_0^1 \frac{dv'}{v'} c(v,v')R(v',u,x)$$

$$+ \int_0^1 c(-v',-u) \frac{dv'}{v'} Q(v,v',x)$$

$$+ \frac{1}{2}\int_0^1\int_0^1 \frac{dv''}{v''}\frac{dv'}{v'} c(-v',v'')R(v'',u,x)Q(v,v',x).$$

$$(7.40)$$

Assuming that the linear Cauchy system for the function Q
in Eqs. (7.40) and (7.37) has a unique solution and keeping
in mind the Cauchy system for the function R, it is seen
that Q = R, or

$$R(v,u,x) = 4\int_0^x e^{-(x-y)/v} J(y,v,x,u) dy ,$$
$$x\geq 0; \ 0\leq v,u,\leq 1 . \quad (7.41)$$

Next introduce the function M by the equation

$$M(t,v,x,u) = \frac{1}{2} c(v,-u) e^{-(x-t)/u}$$

$$+ \int_t^x \int_0^1 \frac{dv'}{v'} c(v,-v') e^{-(y-t)/v'} J(y,-v',x,u) dy$$

$$+ \int_0^t \int_0^1 \frac{dv'}{v'} c(v,v') e^{-(t-y)/v'} J(y,+v',x,u) dy ,$$

$$0 \le t \le x; \quad 0 \le u \le 1; \quad -1 \le v \le +1 . \quad (7.42)$$

At $x = t$ we find that

$$M(t,v,t,u) = \frac{1}{2} c(v,-u) + \int_0^t \int_0^1 \frac{dv'}{v'} c(v,v') e^{-(t-y)/v'}$$

$$J(y,v',t,u) dy$$

$$= \frac{1}{2} c(v,-u) + \int_0^1 \frac{dv'}{v'} c(v,v') \int_0^t e^{-(t-y)/v'}$$

$$J(y,v',t,u) dy$$

$$= \frac{1}{2} c(v,-u) + \frac{1}{4} \int_0^1 \frac{dv'}{v'} c(v,v') Q(v',u,t)$$

$$= \frac{1}{2} c(v,-u) + \frac{1}{4} \int_0^1 \frac{dv'}{v'} c(v,v') R(v',u,t)$$

$$= J(t,v,t,u) . \quad (7.43)$$

In addition, differentiation of both sides of Eq. (7.42) with respect to $x$ shows that

$$M_x(t,v,x,u) = -\frac{1}{2u}\,c(v,-u)e^{-(x-t)/u}$$

$$+ \int_0^1 \frac{dv'}{v'}\,c(v,-v')e^{-(x-t)/v'}J(x,-v',x,u)\,dy$$

$$+ \int_t^x\int_0^1 \frac{dv'}{v'}\,c(v,-v')e^{-(y-t)/v'}\,dy$$

$$\left\{ -\frac{1}{u}\,J(y,-v',x,u) + 2\int_0^1 J(x,-v'',x,u) \right.$$

$$\left. J(y,-v',x,v'')\,\frac{dv''}{v''} \right\}$$

$$+ \int_0^t\int_0^1 \frac{dv'}{v'}\,c(v,v')\,e^{-(t-y)/v'}\,dy$$

$$\left\{ -\frac{1}{u}\,J(y,v',x,u) + 2\int_0^1 J(x,-v'',x,u) \right.$$

$$\left. J(y,v',x,v'')\,\frac{dv''}{v''} \right\}. \qquad (7.44)$$

By collecting terms, this equation becomes

$$M_x(t,v,x,u) = -\frac{1}{u}\,M(t,v,x,u)$$

$$+ 2\int_0^1 \frac{dv'}{v'}\,J(x,-v',x,u)M(t,v,x,v')\,,$$

$$x \geq t \;. \qquad (7.45)$$

From our uniqueness assumption it follows that

$$M(t,v,x,u) \equiv J(t,v,x,u)\,, \qquad\qquad x \geq t\,, \qquad (7.46)$$

which is precisely the family of integral equations in (7.27).

Rational Function Approximation. We wish to calculate reflection functions for several strongly peaked forward scattering diagrams. A simple rational function that exhibits this property is

$$p(\cos\,\theta) = \frac{k}{b - \cos\,\theta} \; , \qquad\qquad (7.47)$$

where  b  is a number slightly greater than  1,  and  k  is
a positive constant chosen to satisfy the normalization condition of the phase function.  The closer  b  is to unity, the
stronger is the forward scattering.  This phase function, as
a function of two polar angles and an azimuth angle is,

$$p(v,u,\phi) = \frac{k}{b - [uv + (1-u^2)^{1/2} (1-v^2)^{1/2} \cos\,\phi\,]} \; .$$
$$\qquad\qquad (7.48)$$

By choosing  $p(\cos\,\phi)$  to be a rational function, we know
that we can write the analytical expression for  $c(v,u)$  by
evaluating the integral which appears on the right-hand side
of equation,

$$c(v,u) = \frac{\lambda k}{4\pi} \int_0^{2\pi} \left\{ b - uv - (1-u^2)^{1/2} (1-v^2)^{1/2} \cos\,\phi \right\}^{-1} d\phi \; .$$
$$\qquad\qquad (7.49)$$

The integral has the form

$$\int_0^{2\pi} \frac{d\phi}{\beta + B \cos\,\phi} \qquad\qquad (7.50)$$

where

$$\beta = b - uv \; ,$$
$$B = (1-u^2)^{1/2} (1-v^2)^{1/2} \; . \qquad\qquad (7.51)$$

Apply the integration formula

$$\int_0^{2\pi} \frac{dx}{1 + a \cos x} = \frac{2\pi}{(1-a^2)^{1/2}} \ , \qquad a^2 < 1 \ . \qquad (7.52)$$

The integral is then seen to be

$$\int_0^{2\pi} \frac{d\phi}{\beta + B \cos \phi} = 2\pi \ [(b-A)^2 - B^2]^{-1/2} \ . \qquad (7.53)$$

$$(A = uv)$$

Then   $c(v,u)$   is

$$c(v,u) = 0.5 \ \lambda k \ [(b-uv)^2 - (1-u^2)(1-v^2)]^{-1/2} \ .$$

$$(7.54)$$

The constant   $k$   must be specified so as to satisfy the
normalization conditions.   Let

$$z = \cos \phi \ . \qquad (7.55)$$

The condition is written

$$\frac{1}{4\pi} \int_{-1}^{+1} p(z) \ 2\pi \ dz = 1 \qquad (7.56)$$

or

$$\frac{k}{2} \int_{-1}^{+1} \frac{dz}{b-z} = 1 \ . \qquad (7.57)$$

Then   $k$   must have the value

$$k = 2 \left( \log \frac{b+1}{b-1} \right)^{-1} . \tag{7.58}$$

The fraction of the energy which is scattered into the forward hemisphere following a single scattering event is

$$f = 2\pi \int_0^1 \frac{p(\cos \phi)}{4\pi} \left| d(\cos \phi) \right| \tag{7.59}$$

which, in this case, is

$$f = \frac{k}{2} \log \frac{b}{b-1} . \tag{7.60}$$

In Table 7.1 are found the fraction , f, of scattering into the forward hemisphere, and the ratio of forward to backward scattering, $p(\cos 0^{\circ})/p(\cos 180^{\circ})$, for each of six values of the parameter b.

Table 7.1

Forward Scattering Parameters

| b | f | $p(\cos 0^{\circ})/p(\cos 180^{\circ})$ |
|---|---|---|
| 100.0 | 0.500 | 1.00 |
| 2.0 | 0.631 | 3.59 |
| 1.5 | 0.683 | 5.00 |
| 1.1 | 0.788 | 21.0 |
| 1.01 | 0.870 | 201. |
| 1.001 | 0.909 | 2008. |

   Numerical Method.  The initial value method for
computing reflected intensities is based on Gaussian
quadrature and the method of lines.  Let $\{z_i\}$ and $\{w_i\}$,
$i = 1,2,\ldots,N$, denote abscissas and weights respectively
for Gaussian quadrature of order $N$.  Let

$$R_{ij}(x) = R(z_i, z_j, x) , \qquad i,j=1,2,\ldots,N . \qquad (7.61)$$

For a choice of $L$ direction cosines $\{a_\ell\}$, $\ell = 1,2,\ldots,L$,
we define the functions

$$A_{i\ell}(x) = R(z_i, a_\ell, x) , \qquad\qquad (7.62)$$

$$B_{k\ell}(x) = R(a_k, a_\ell, x) , \qquad\qquad (7.63)$$

$$i - 1,2,\ldots,N; \; \ell,k = 1,2,\ldots,L .$$

The functions $R_{ij}(x)$ satisfy the closed system,

$$R'_{ij} = - (z_i^{-1} + z_j^{-1}) R_{ij} + 2 c(z_i, - z_j)$$

$$+ \sum_{m=1}^{N} c(z_i, z_m) R_{mj} w_m/z_m$$

$$+ 2 \sum_{m=1}^{N} R_{im} w_m/z_m \left\{ \frac{1}{2} c(-z_m, -z_j) \right.$$

$$+ \frac{1}{4} \sum_{n=1}^{N} c(-z_m, z_n) R_{nj} w_n/z_n \Big\} , \qquad (7.64)$$

$$R_{ij}(0) = 0 , \qquad\qquad (7.65)$$

and the $A_{i\ell}(x)$ and $\beta_{k\ell}(x)$ satisfy similar equations which can be adjoined to the basic system above. By so introducing the functions $A_{i\ell}$ and $\beta_{k\ell}$, we may compute reflected intensities for arbitrary pairs of incident and reflected angles,

$$r(a_k, a_\ell, x) = B_{k\ell}(x)/4 \, a_k \, . \tag{7.66}$$

We may also wish to change the value of N while still being able to compare intensities for arbitrary angles. Then this is important for checking purposes and for estimating accuracies.

On the other hand, we can interpolate at desired angles by expansion in orthogonal polynomials. See Ref. 1.

Numerical Results. Some curves for reflection functions are shown in Ref. 1. The value of N and the integration step sizes used to calculate the reflection functions with reasonable accuracy are given in Table 7.2. We also list the approximate number of significant figures obtained. These estimates are made by varying N and $\Delta x$ (the step size of integration) and comparing the numerical values of r. Attempts are made to calculate the reflection function for the extremely peaked phase function for b = 1.001. As can be seen from this table, highly accurate results are not obtained for this case, even with a large number of quadrature points and a tiny integration step size. The computing times are 1 min with N = 7 and $\Delta x = 0.01$.

Table 7.2

Accuracy of Evaluation of  r  by Integration

| b | N | Δx | Approximate Number of Significant Figures |
|---|---|-----|---|
| 1000. | 7 | 0.01 | 6 |
| 1.5 | 7 | 0.01 | 4-5 |
| 1.1 | 7 | 0.01 | 4-5 |
| 1.01 | 9 | 0.005 | 2-3 |
| 1.001 | 15 | 0.001 | 1 |

We apply the second interpolation scheme, expansion in orthogonal polynomials, to obtain values of  r  at evenly spaced input cosine values, first when the output cosines are  v = 0.025, 0.5, and 0.975, and again for evenly spaced output cosines when the input cosines are  u = 0.1  and  1.0. These evaluations are compared with those obtained by integrating additional equations for  r, in Table 7.3 for the case in which  b = 1.1, $\lambda$ = 1, and  x = 1.  In the expansion calculation, the coefficients  $\beta_i(v)$ (see Ref. 1) were computed in double precision arithmetic (about 15 to 16 decimal figures), for  v (or u) = 0.1, 0.2, ...,1.0,  and  i = 1,2,...,7.  We first compute the  r  functions using about eight figures of accuracy in  $\beta_i(v)$.  Then we repeat the calculations after truncating  $\beta_i(v)$  after the sixth decimal place.  The  r  functions using six and eight figures are in agreement to about six places.

Table 7.3

Reflection Functions by Two Interpolation Methods
for the Case b = 1.1, λ = 1, x = 1

| u | v = 0.025 | | v = 0.500 | | v = 0.975 | |
|---|---|---|---|---|---|---|
| | Integ. | Expans. | Integ. | Expans. | Integ. | Expans. |
| 0.1 | .3498 | .344 | .0733 | .0734 | .03178 | .03176 |
| 0.2 | .3985 | .403 | .1335 | .1334 | .06330 | .06329 |
| 0.3 | .4054 | .405 | .1774 | .1774 | .0894 | .0894 |
| 0.4 | .3957 | .394 | .2069 | .2070 | .1087 | .1087 |
| 0.5 | .3791 | .379 | .2253 | .2253 | .1220 | .1220 |
| 0.6 | .3601 | .362 | .2357 | .2356 | .1306 | .1306 |
| 0.7 | .3408 | .341 | .2406 | .2406 | .1358 | .1358 |
| 0.8 | .3222 | .321 | .2416 | .2417 | .1384 | .1385 |
| 0.9 | .3049 | .306 | .2401 | .2400 | .1392 | .1392 |
| 1.0 | .2887 | .286 | .2368 | .2370 | .1388 | .1389 |

| v | u = 0.1 | | u = 1.0 | |
|---|---|---|---|---|
| | Integ. | Expans. | Integ. | Expans. |
| 0.1 | .2360 | .238 | .3063 | .3062 |
| 0.2 | .1612 | .160 | .3059 | .3057 |
| 0.3 | .1192 | .119 | .2885 | .2885 |
| 0.4 | .0921 | .0930 | .2634 | .2636 |
| 0.5 | .0733 | .0731 | .2368 | .2368 |
| 0.6 | .0597 | .0591 | .2114 | .2113 |
| 0.7 | .0495 | .0495 | .1885 | .1885 |
| 0.8 | .0417 | .0422 | .1683 | .1683 |
| 0.9 | .0355 | .0352 | .1505 | .1504 |
| 1.0 | .0306 | .0320 | .1350 | .1353 |

## 7.3     DISCUSSION

The theoretical and numerical treatment of multiple
scattering involving highly anisotropic elementary acts of
scattering is still in a rudimentary state.  We have chosen
to attack some problems for which there is no azimuth
dependence of the diffuse radiation field though the local
scattering is anisotropic.  This is half way between the
desired case  and the simple isotropic case.  To simulate
the great forward scattering lobes suggested by Deirmendjian
and others, we suggest using a simple rational function,
for both its analytical and its computational convenience.
More realistic approximations will require the use of higher
polynomials in both the numerator and denominator.  In
these cases it may be necessary to evaluate the function
$c(v,u)$ numerically.  The function  $c(v,u)$  would have a
sharp peak along  $v = u$  that could cause difficulty in
evaluating the integrals in Eqs. (7.32) and (7.33).

The treatment of a variety of inverse problems for
anisotropic scattering, i.e., the estimation of the local
phase function based on multiple scattering measurements,
now appears feasible.  A start is made in a later section.

Extensions to the case of isotropic sources,
reflecting surfaces, and so forth are readily carried out.


## 7.4     EXPANSION IN LEGENDRE POLYHOMIALS

We return to a consideration of the initial value
problem for the reflection function,  $R(v,\phi,u,x)$   as stated

in Eqs. (7.18) and (7.19). We reduce the number of variables from four to three thus lightening the computational
task.

Assume that the phase function may be expanded in a
series of Legendre polynomials consisting of M+1 terms,
where M is finite and about 20 or less,

$$p(\cos \phi) = \sum_{m=0}^{M} c_m P_m (\cos \phi) . \qquad (7.67)$$

Expand the reflection function in a similar manner,

$$R(v,\phi,u,x) = \sum_{m=0}^{M} R^m(v,u,x) \cos m\phi . \qquad (7.68)$$

Substitution of the above expression in Eqs. (7.18) and (7.19)
and application of the addition theorem for Legendre functions lead to the initial value problem for each
$R^m = R^m(v,u,x)$ ,

$$\frac{\partial}{\partial x} R^m(v,u,x) = - (v^{-1} + u^{-1}) R^m(v,u,x)$$

$$+ \lambda (2-\delta_{0m}) \sum_{k=m}^{M} (-1)^{k+m} c_k \frac{(k-m)!}{(k+m)!} \psi_k^m(v)\psi_k^m(u) ,$$

$$(7.69)$$

$$R^m(v,u,0) = 0 , \qquad (7.70)$$

for m = 0, 1,...,M. There are M+1 such problems. In
Eq. (7.69),

$$\psi_k^m(w) = P_k^m(w) + \frac{(-1)^{k+m}}{2(2-\delta_{0m})} \int_0^1 R^m(w,z,x) P_k^m(z) \frac{dz}{z} ,$$

$$m=0,1,2,\ldots,M; \quad k=m,m+1,\ldots,M , \qquad (7.71)$$

where $\delta_{0m}$ is the Kronecker delta function ($\delta_{0m} = 1$ if $m = 0$; otherwise $\delta_{0m} = 0$), and $P_k^m(w)$ is an associated Legendre function.

Gaussian quadrature of order $N$ approximates the integral which appears in Eq. (7.71). Let

$$R_{ij}^m(x) = R^m(z_i,z_j,x) , \qquad (7.72)$$

$$m = 0,1,2,\ldots,M; \quad i,j-1,2,\ldots,N ,$$

where $\{ z_i \}$ are the quadrature points. Denoting the weights $\{ w_i \}$, we write Eq. (7.71) as

$$\psi_{ki}^m = P_k^m(z_i) + \frac{(-1)^{k+m}}{2(2-\delta_{0m})} \sum_{j=1}^N R_{ij}^m P_k^m(z_i) \frac{w_j}{z_j} . \qquad (7.73)$$

Then Eq. (7.69) becomes

$$\frac{d}{dx} R_{ij}^m(x) = - (z_i^{-1} + z_j^{-1}) R_{ij}^m$$

$$+ \lambda(2-\delta_{0m}) \sum_{k=m}^M (-1)^{k+m} \frac{(k-m)!}{(k+m)!} c_k \psi_{ki}^m \psi_{kj}^m , \qquad (7.74)$$

and the initial conditions,

$$R_{ij}^m(0) = 0 ,$$

$$m = 0,1,\ldots,M; \quad i,j=1,2,\ldots,N . \qquad (7.75)$$

There are $(M+1)N^2$ equations implied in Eqs. (7.74).
However, due to the symmetry property, $R^m_{ij} = R^m_{ji}$ , only
$(M+1)N(N+1)/2$ equations need be integrated. Reflected intensities are calculated from initial value problems for
ordinary differential equations. These problems have been
found to be stable and well-suited for high speed computers.
The value of M however should not be larger than about 20
in order that the integrals be evaluated accurately. After
all, a Legendre function of order 20 oscillates very rapidly.
This means that additional measures must be taken to insure
accuracy for highly peaked phase functions.

The authors have a Fortran code for computing
reflected intensities. For the case $x = 0.2$, $p(\cos \phi) = 1$
$+ \cos \phi$, $\lambda = 1$, $N = 7$, $M = 1$, and for seven azimuth angles,
$0^o$, $30^o$,...,$180^o$, the computing time is several seconds on an
IBM 7044. An integration step size of 0.01 is used with a
fourth order predictor-corrector method. Additional numerical
calculations are consistent with those of Ref. 3.


7.5    ESTIMATION OF PHASE FUNCTIONS BASED ON MULTIPLE
       SCATTERING DATA


It is of interest to estimate the phase function in
a terrestrial or planetary atmosphere based on observations
made from satellites, and from the phase function deduce
the nature of the scattering particles. If the phase function is not too highly peaked, the expansion in Legendre
polynomials is appropriate. Then the initial value method,
together with quasilinearization, may enable us to attack
such inverse problems in a systematic way. We discuss an

initial study for a rather simple example. The calculations
can quickly become sizable for even moderate values of M.
We consider the case in which the phase function is

$$p(\cos \phi) = 1 + a \cos \phi \ , \qquad\qquad M = 1 \ , \qquad (7.76)$$

and the true value of a is 1.0. The thickness is x = 0.2,
the scattering is conservative, $\lambda = 1$. There are 343 pieces
of data: reflected intensities in 49 directions for each of
seven incident directions. In some experiments, accurate
observations are used. In others, the observations contain
various amounts of Gaussian noise.

Several computational experiments are performed.
In three controlled experiments without any errors in the
observations, but with wrong initial estimates (a = 0.8,
0.5, and 0.0), the refined estimate a = 0.99999 is obtained.
Note that setting a = 0 implies isotropic scattering as
the initial estimate.

In the second series of experiments, errors of 0.5,
1,2,5, and 10 percent are introduced into the observations
of r, and estimates of the parameter a are obtained which
are all less than approximately 1 percent in discrepancy.
In the experiment with 10 percent error and with isotropic
scattering as the initial estimate, the sequence of
approximations of the coefficient a is as follows:

$$a^0 = 0.0 \ ,$$
$$a^1 = 1.087,$$
$$a^2 = 1.014 \ ,$$
$$a^3 = 1.014 \ . \qquad\qquad\qquad\qquad (7.75)$$

Thus, even with large errors in the measurements and with a completely wrong initial estimate of the coefficient, the estimate of the coefficient is refined in only two steps to a discrepancy of only 1.4 percent. This calculation required less than 1.5 min of computing time on an IBM 7044.

We have shown that it is possible to determine local scattering properties based on measurements of multiple scattered radiation. The ability of modern digital computers to integrate large systems of ordinary differential equations, and a new mathematical technique, quasilinearization, play essential roles. Subsequent work will deal with sharply peaked forward scattering, polarization, and with inhomogeneous spherical atmospheres.

The method readily extends to inhomogeneous media. For example, suppose that the phase function is dependent on optical height,

$$p(\cos \theta, t) = 1 + a(t) \cos \theta , \qquad (7.76)$$

where

$$a(t) = a_o + a_1 t + a_2 t^2 \qquad (7.77)$$

and $a_o, a_1$ and $a_2$ are unknown constants. Then the expansion becomes

$$p(\cos \phi, t) = \sum_{m=0}^{1} c_m(t) \ P_m(\cos \theta) \qquad (7.78)$$

$$c_o(t) = 1 , \qquad (7.79)$$

$$c_1(t) = a_o + a_1 t + a_2 t^2 .$$

(7.80)

The procedure is similar to that discussed.

An alternative model is to describe the coefficient $a(t)$ as the solution of a differential equation,

$$\dot{a} = g(t, b_o, b_1, \ldots, b_n) ,$$

(7.81)

where $b_o, b_1, \ldots, b_n$ are unknown constants. The solution may be a polynomial, exponential trigonometric function, etc. depending on the coefficients.

The functions b and h provide yet another method of reduction of the anisotropic scattering problem to functions of fewer variables. This new theory will be presented in a subsequent book.

## EXERCISES

1.  Derive the Cauchy system for the internal intensity
    function $I(t,v,\phi,x,u)$.

2.  Validate the Cauchy system for the source function.

3.  Derive the Cauchy system for the functions $J$ and
    $R$ in the case of axially symmetric fields.

4.  Derive the Cauchy system for $J$, $R$ and $I$ for the
    case of inhomogeneous anisotropic medium with bottom
    Lambert reflector.  Compare with Ref. 2.

## REFERENCES

1. H. Kagiwada and R. Kalaba, "Multiple Anisotropic Scattering in Slabs with Axially Symmetric Fields," The Rand Corporation, RM-5245-PR, February 1967.

2. R. Bellman, H. Kagiwada, R. Kalaba and S. Ueno, "Chandrasekhar's Planetary Problem with Integral Sources," *Icarus*, v. 7, no. 3, 1967, pp. 365-371.

3. E. Feigel'son, M. Maldevich, S. Kogan, T. Koronatova, K. Glazova and M. Kuznetsova, "Calculation of the Brightness of Light in the Case of Anisotropic Scattering, Part 1," *Trudy Inst. Fiz. Atm. Akad. Nauk SSSR*, No. 1, Moscow, 1958.

## BIBLIOGRAPHY

Abhyankar, K. D. and Fymat, A. L., 1971, *Astrophys. J. Suppl.*
   Vol. 23, 35.

Ambartsumian, V. A., 1958, *Theoretical Astrophysics*, Pergamon
   Press, New York, 645 pp.

Bellman, R., Kagiwada, H., Kalaba, R., and Ueno, S., 1967,
   *Icarus* 7, 365.

Bellman, R.E., Kalaba, R.E. and Prestrud, M.C., 1963, *Invariant Imbedding and Radiative Transfer in Slabs of
   of Finite Thickness*, American Elsevier, New York, 346 pp.

Busbridge, I. W., 1960, *The Mathematics of Radiative Transfer*,
   Cambridge Univ. Press, Cambridge, 143 pp.

Carlstedt, J. L. and Mullikin, T. W., 1966, *Astrophys. J.
   Suppl.*, 12, 449.

Case, K. M. and Zweifel, P. F., 1967, *Linear Transport Theory*,
   Addison-Wesley, Reading, Mass., 342 pp.

Coulson, K. L., Dave, J. V., and Sekera, Z., 1960, *Tables
   Related to Radiation Emerging from a Planetary Atmosphere
   with Rayleigh Scattering*, Univ. of Calif. Press, Berkeley,
   584 pp.

Davison, B., 1958, *Neutron Transport Theory*, Oxford Univ.
   Press, London, 450 pp.

Deirmendjian, D., 1969, *Electromagnetic Scattering on
   Spherical Polydispersions*, American Elsevier, New York,
   290 pp.

Fymat, A.L.,1975, *Appl. Math. and Comput.* 1, 131.

Grant, I.P. and Hunt, G.E., 1969, *Proc. Roy. Soc. London*,
   A 313, 183.

Hansen, J.E. and L. D. Travis, 1974, *Light Scattering in
   Planetary Atmospheres*, Space Sciences Reviews 16, 527.

Horak, H.G., 1950, *Astrophys. J.*, 112, 445.

Irvine, W.M., 1965, *Astrophys. J.*, 142, 1563.

Kattawar, G. W., Plass, G. N., and Guinn, J. A., 1973, *J. Phys. Ocean*, 3, 353.

Lenoble, J., 1970, *J. Quant. Spectr. Radiat. Transfer*, 10, 533.

Lorenz, L., 1898, *Oeuvres Scientifiques*, Vol. 1, Lehman and Stage, Copenhagen, p. 405.

Marchuk, G. I. and Mikhailov, G.A., 1967a, *Izv. Atmospheric Ocean. Phys.* 3, 258.

Mie, G., 1908, *Ann. Phys.*, 25, 377.

Mingle, J.D., 1973, *The Invariant Imbedding Theory of Nuclear Transport*, American Elsevier, New York, 131 pp.

Pahor, S. and Kuscer, I., 1966, *Astrophys. J.*, 143, 888.

Preisendorfer, R.W., 1968, *J. Quant. Spectros. Radiat. Transfer*, 8, 325.

Rayleigh, Lord (J. W. Strutt), 1871, *Phil. Mag.*, 41, 107, 274 and 477.

Schuster, A., 1905, *Astrophys. J.*, 21, 1.

Schwarzschild, K., 1906, Gottinger Nachrichten, 41.

Sobolev, V.V., 1974, *Light Scattering in Planetary Atmospheres*, Pergammon Press, New York (kn press).

Stibbs, D.W.N. and Weir, R.E., 1959, *Monthly Notices Roy. Astron. Soc.*, 119, 512.

Sweigart, A.V., 1970, *Astrophys. J. Suppl.*, 22, 1.

Uesugi, A. and Irvine, W.M., 1970, *Astrophys. J.*, 159, 127.

van de Hulst, H.C., 1957, *Light Scattering by Small Particles*, Wiley, New York, 470 pp.

Wark, D.Q. and Fleming, H.E., 1966, *Monthly Weather Review*,
Vol. 94, p. 351.

Wing, G. M., 1962, *An Introduction to Transport Theory*,
John Wiley, New York, 169 pp.

APPENDICES

## APPENDIX A

## POINTS, WEIGHTS AND ANGLES FOR GAUSSIAN QUADRATURE WITH $N = 7$

$$\int_0^1 f(x)\,dx \cong \sum_{i=1}^{N} f(z_i) w_i$$

| $i$ | Points $z_i$ | Weights $w_i$ | Angles (deg) $\cos^{-1} z_i$ |
|---|---|---|---|
| 1 | .025446046 | .064742484 | 88.5419 |
| 2 | .12923441 | .13985269 | 82.5746 |
| 3 | .29707742 | .19091502 | 72.7178 |
| 4 | .50000000 | .20897958 | 60.0000 |
| 5 | .70292258 | .19091502 | 45.3380 |
| 6 | .87076559 | .13985269 | 29.4523 |
| 7 | .97455396 | .064742484 | 12.9531 |

## APPENDIX B

### FORTRAN PROGRAM FOR b AND h FUNCTIONS

In this Appendix is listed a FORTRAN program for the computation of b, h and $\Phi$ functions. The program was written in FORTRAN IV for an IBM 7044.

Some of the input data are the following:

| | |
|---|---|
| N | Number points for gaussian quadrature |
| RT(I),I = 1,N | Quadrature points $z_1^{(N)}, z_2^{(N)}, \ldots, z_N^{(N)}$ |
| WT(I),I = 1,N | Quadrature weights $w_1^{(N)}, w_2^{(N)}, \ldots, w_N^{(N)}$ |
| DELTA | Step size of integration of ordinary differential equations ($\Delta x$) |
| ALAM | Albedo for single scattering ($\lambda$) |

The superscripts (N) will henceforth in this Appendix be omitted from the symbols for quadrature points and weights.

The integration of a system of NEQ ordinary differential equations is carried out by means of two subroutines INTS and INTM (not listed here) for integration of ordinary differential equations. Subroutine INTS is called once to set up the initial conditions and some auxiliary quantities. The third argument in INTS is put equal to 2, indicating the choice of the Adams-Moulton fourth-order predictor-corrector method of integration. The first argument, T, is an array dimensioned (12*NEQ + 3) or greater. It contains the following variables:

| T(2) | Current value of independent variable, x |
|------|------------------------------------------|
| T(3) | Step size of integration, x |
| T(4),...,T(NEQ+3) | Current values of the NEQ dependent variables |
| T(NEQ+4),...,T(2*NEQ+3) | Current values of the NEQ derivatives |
| All other T | Auxiliary data |

A call to subroutine INTM when $x = x_i$ produces the solution of the differential equation (as well as the derivatives) at the next point, $x = x_i + x$. Note that the right-hand sides of the differential equations are coded, in terms of the T array, in subroutine DAUX of each program. The subroutine DAUX is called upon by INTS and INTM.

The listed program computes the X, Y, J, b, h, $\Phi$ r, and t functions of isotropic scattering with albedo $\lambda$. These functions are evaluated for thicknesses $0 \leq x \leq x_M$, for directions whose cosines are $z_1, z_2, \ldots, z_N$. The functions J, b, h, and $\Phi$ are evaluated at various altitudes $0, t_1, t_2, \ldots$ The b, h and $\Phi$ functions are the only quantities printed by this program. The purpose of this program is to tabulate the b, h and $\Phi$ functions from which all of the other functions, X, Y, J, r, t, and I, may be readily evaluated. The exact and approximate initial value problems are discussed and formulas for X, Y, J, etc. in terms of b and h are given in Chapter 3.

Let us use the notation

$$X_j(x) = X(z_j,x) , \qquad Y_j(x) = Y(z_j,x) ,$$

$$T_{kj}(x) = T(z_k,z_j,x) , \qquad J_{ij}(x) = J(t_i,z_j,x) ,$$

$$b_{ik}(x) = b(t_i,-z_k,x) , \qquad h_{ik}(x) = h(t_i,-z_k,x) ,$$

$$\Phi_i(x) = \Phi(t_i,x) , \qquad \Phi_0(x) = \Phi(0,x) ,$$

$$r_{kj}(x) = r(z_k,z_j,x) , \qquad t_{kj}(x) = t(z_k,z_j,x) .$$

Note that $b_{ik}$ and $h_{ik}$ are b and h functions in downwelling directions.

The program listed here solves the initial value problem with differential equations

$$X_j' = Y_j \Phi_0 \qquad\qquad x{\geq}0 , \qquad (B.1)$$

$$Y_j' = -z_j^{-1} Y_j + X_j \Phi_0 \qquad x{\geq}0 , \qquad (B.2)$$

$$T_{jj}' = -z_j^{-1} T_{jj} + \lambda X_j Y_j \qquad x{\geq}0 , \qquad (B.3)$$

$$J_{ij}' = -z_j^{-1} J_{ij} + (\lambda/4) X_j \Phi_i \qquad x{\geq}t_i , \qquad (B.4)$$

$$b_{ik}' = -z_i^{-1} [b_{ik} - \Phi_i] + h_{ik}\Phi_0 , \quad x{\geq}t_i , \quad (B.5)$$

$$h_{ik}' = b_{ik}\Phi_0 \qquad\qquad x{\geq}t_i , \qquad (B.6)$$

$$j = 1,2,\ldots,N; \quad k = 1,2,\ldots,N$$

and initial conditions

$$X_j(0) = 1 \ , \tag{B.7}$$

$$Y_j(0) = 1 \ , \tag{B.8}$$

$$T_{jj}(0) = 0 \ , \tag{B.9}$$

$$J_{ij}(t_i) = (\lambda/4) \ X_j(t_i) \ , \tag{B.10}$$

$$b_{ik}(t_i) = z_k^{-1} \ , \tag{B.11}$$

$$h_{ik}(t_i) = 0 \ , \tag{B.12}$$

where

$$\Phi_0 = (\lambda/2) \sum_{m=1}^{N} Y_m w_m/z_m \ , \tag{B.13}$$

$$\Phi_i = 2 \sum_{m=1}^{N} J_{im} w_m/z_m \ . \tag{B.14}$$

The altitudes $t_i$ are introduced at specific points in the program as

$$t_1 = x_1 \quad \text{when} \quad x = x_1 ,$$

$$t_2 = x_2 \quad \text{when} \quad x = x_2 ,$$

$$\ldots$$

$$t_M = x_M \quad \text{when} \quad x = x_M \ .$$

The 3N differential equations (B.1) - (B.3) for $X_j, Y_j$ and $T_{jj}$ $(j = 1, 2, ..., N)$ are numerically integrated from $x = 0$, with the initial conditions of Eqs. (B.7) - (B.9), to $x = x_1$. At this point the initial conditions of Eqs. (B.1) - (B.12) are imposed on $J_{ij}$, $b_{ik}$ and $h_{ik}$ $(k = 1, 2, ..., N)$. The 3N differential eqs. (B.4) - (B.6) with $i = 1$ are adjoined and the entire system is integrated for $x$ increasing until $x = x_2$. When $x = x_2$, a new set of initial conditions is imposed on the new functions $J_{2j}, b_{2k}$, and $h_{2k}$ and the additional differential Eqs. (B.4) - (B.6) with $i = 2$ are adjoined. The ever-enlarging system is integrated in this manner until $x$ reaches the maximum value, $x_M$.

The b, h and $\Phi$ functions are printed out when $x$ attains the balues $x_1, x_2, ..., x_M$. The b and h functions at the bottom $(t = 0)$ are evaluated, in downwelling directions, as

$$b(0, -z_k, x) = 2 \sum_{m=1}^{N} t_{km}(x) + z_k^{-1} \exp(-x/z_k) , \qquad (B.15)$$

$$h(0, -z_k, x) = 2 \sum_{m=1}^{N} r_{km}(x) , \qquad (B.16)$$

where

$$r_{kj} = \frac{\lambda}{4} \frac{z_i}{z_k + z_j} \left\{ X_k X_j - Y_k Y_k \right\} . \qquad (B.17)$$

$$t_{kj} = \frac{\lambda}{4} \frac{z_j}{z_k - z_j} \left\{ Y_k X_j - X_k Y_j \right\} , \qquad k \neq j , \qquad (B.18)$$

$$t_{jj} = T_{jj}/4z_j .$$
(B.19)

In upwelling directions, they are

$$b(0, + z_k, x) = 0 ,$$
(B.20)

$$h(0, +z_k, x) = z_k^{-1} .$$
(B.21)

The b and h functions at a general altitude $t_i$, in upwelling directions, are determined from computed values of b and h in downwelling directions according to the formulas

$$b(t_i, + z_k, x) = h(x-t_i, -z_k, x) ,$$
(B.22)

$$h(t_i, + z_k, x) = k(x-t_i, -z_k, x) .$$
(B.23)

The FORTRAN variables K1MAX, DELTA, and MMAX determine the maximum value of the thickness as well as the values of x for which b, h, and $\Phi$ are output. DELTA is the step size of integration, and MMAX = M - 1 where M is as before. The values of x at printout are

$$x_1 = K1MAX*DELTA,$$
$$x_2 = x_1 + KMAX*DELTA ,$$
$$x_3 = x_2 + KMAX*DELTA ,$$
$$...$$
$$x_M = x_1 + MMAX*KMAX*DELTA .$$

Sample input data are shown at the end of the program listing. The tables of b, h and $\Phi$ in Appendix C were generated by this program.

```
$JOB           2890,BHFUNS,K0160,05,200,100,C      PAL 77
$IBJOB         MAP
$IBFTC MAIN
C              SOURCE FUNCTION S(I,J)
C              B AND H FUNCTIONS B(I,K), H(I,K)
C              I REFERS TO I-TH ALTITUDE
C              J REFERS TO J-TH DIRECTION COSINE OF INCIDENCE *
C              K REFERS TO K-TH DIRECTION COSINE OF PROPAGATION
C              MMAX=MAX. NO. OF SLABS - 1
C              KMAX=NO. OF INTEGRATIONS PER INCREASE IN THICKNESS
C              K1MAX=NO. OF INTEGRATIONS TO FIRST SLAB
C
       COMMON T(11119),N,MMAX,KMAX,K1MAX,RT(15),WT(15),DELTA,ALAM,THICK,
      1 NP,N2,NADD,AL2,AL4,WR(15),NEQ,TR(15),ALT,B(20,30),H(20,30),PO,
      2 P(30),BO(30),HO(30),MP,IFLAG,X(15),Y(15),REF(15,15),TRA(15,15)
      3 ,S(20,15),SO(15),DEP,DALT,IPRNT,NMAX,IO(20)
C
C
   1   READ(5,100)N,MMAX,KMAX,K1MAX
       WRITE(6,92)
       WRITE(6,90)N,MMAX,KMAX,K1MAX
       READ(5,101)(RT(I),I=1,N)
       WRITE(6,91)(RT(I),I=1,N)
       READ(5,101)(WT(I),I=1,N)
       WRITE(6,91)(WT(I),I=1,N)
       RFAD(5,101)DELTA,ALAM,THICK
       WRITE(6,91)DELTA,ALAM,THICK
       NP=N+1
       N2=2*N
       NADD=3*N
       AL2=0.5*ALAM
       AL4=0.25*ALAM
       DO 30 I=1,N
  30   WR(I)=WT(I)/RT(I)
       WRITE(6,91)(WR(I),I=1,N)
C
C              INTEGRATE X, Y, AND T EQS.
C
       NEQ=3*N
       M=0
       MP=M+1
       DO 2 I=1,11119
   2   T(I)=0.0
       T(3)=DELTA
       LL=2*N+3
       DO 7 I=4,LL
   7   T(I)=1.0
       IFLAG=0
       CALL INTS(T,NEQ,2,0,0,0,0,0,0)
C
       DO 8 K=1,K1MAX
   8   CALL INTM
       L=3
       DO 9 I=1,N
       L=L+1
```

```
    9   X(I)=T(L)
        DO 10 I=1,N
        L=L+1
   10   Y(I)=T(L)
        DO 3 I=1,N
        L=L+1
    3   TR(I)=T(L)
        CALL REFTRA
C
C              FIRST SLAB
C
        DO 21 I=1,N
        S(1,I)=AL4*X(I)
   21   SO(I)=AL4*Y(I)
        F=KMAX
        DALT=F*DELTA
C
   43   CALL OUTPUT
C
C
C              MMAX SLABS
C
        DO 50 M=1,MMAX
        MP=M+1
        THICK=THICK + DALT
C
        IFLAG=M
        L=NEQ + 3
        DO 12 I=1,N
        L=L+1
   12   T(L)=S(M,I)
        DO 4 I=NP,N2
        L=L+1
    4   T(L)=B(M,I)
        DO 5 I=NP,N2
        L=L+1
    5   T(L)=H(M,I)
        NEQ = NEQ + NADD
        CALL INTS(T,NEQ,2,0,0,0,0,0,0)
C
        DO 13 K=1,KMAX
   13   CALL INTM
C
        L=3
        DO 14 I=1,N
        L=L+1
   14   X(I)=T(L)
        DO 15 I=1,N
        L=L+1
   15   Y(I)=T(L)
        DO 22 I=1,N
        L=L+1
   22   TR(I)=T(L)
        CALL REFTRA
C
```

```
      DO 16 I=1,N
  16  SO(I)=AL4*Y(I)
      DO 18 I=1,N
  18  S(MP,I)=AL4*X(I)
C
      DO 25 I=1,M
      DO 17 J=1,N
      L=L+1
  17  S(I,J)=T(L)
      DO 23 J=NP,N2
      L=L+1
  23  B(I,J)=T(L)
      DO 24 J=NP,N2
      L=L+1
  24  H(I,J)=T(L)
  25  CONTINUE
C
  49  CALL OUTPUT
C
C
  50  CONTINUE
C
      GO TO 1
 100  FORMAT(12I6)
 101  FORMAT(6E12.8)
  90  FORMAT(12I10)
  91  FORMAT(6F20.8)
  92  FORMAT(1H1)
      END
$IBFTC DAUX
      SUBROUTINE DAUX
      DIMENSION SV(15),BV(30),HV(30)
      COMMON T(11119),N,MMAX,KMAX,K1MAX,RT(15),WT(15),DELTA,ALAM,THICK,
     1 NP,N2,NADD,AL2,AL4,WR(15),NEQ,TR(15),ALT,B(20,30),H(20,30),PO,
     2 P(30),BO(30),HO(30),MP,IFLAG,X(15),Y(15),REF(15,15),TRA(15,15)
     3 ,S(20,15),SO(15),DEP,DALT,IPRNT,NMAX,IO(20)
C
C
      L=3
      DO 1 I=1,N
      L=L+1
   1  X(I)=T(L)
      DO 2 I=1,N
      L=L+1
   2  Y(I)=T(L)
      DO 3 I=1,N
      L=L+1
   3  TR(I)=T(L)
C
      LL=NEQ + 3
      PO=0.0
      DO 4 I=1,N
   4  PO=PO + Y(I)*WR(I)
      PO=PO*AL2
      DO 5 I=1,N
```

```
          LL=LL+1
     5    T(LL)= Y(I)*PO
          DO 6 I=1,N
          LL=LL+1
     6    T(LL)=-Y(I)/RT(I) + X(I)*PO
          DO 7 I=1,N
          LL=LL+1
     7    T(LL)=-TR(I)/RT(I) + ALAM*X(I)*Y(I)
C
          IF(IFLAG)8,8,9
     8    RETURN
C
C                 S, B, AND H
C
     9    DO 19 M=1,IFLAG
C
          DO 10 I=1,N
          L=L+1
    10    SV(I)=T(L)
          DO 11 I=NP,N2
          L=L+1
    11    BV(I)=T(L)
          DO 14 I=NP,N2
          L=L+1
    14    HV(I)=T(L)
          SUM=0.0
          DO 15 K=1,N
    15    SUM=SUM + SV(K)*WR(K)
          P(M)=2.0*SUM
          DO 16 I=1,N
          LL=LL+1
    16    T(LL)=-SV(I)/RT(I) + AL4*X(I)*P(M)
          DO 18 I=1,N
          NI=N+I
          LL=LL+1
    18    T(LL)=-(BV(NI)-P(M))/RT(I) + PO*HV(NI)
          DO 19 I=NP,N2
          LL=LL+1
    19    T(LL)=PO*BV(I)
C
          RETURN
          END
$IBFTC REFTRA
          SUBROUTINE REFTRA
          COMMON T(11119),N,MMAX,KMAX,K1MAX,RT(15),WT(15),DELTA,ALAM,THICK,
         1 NP,N2,NADD,AL2,AL4,WR(15),NEQ,TR(15),ALT,B(20,30),H(20,30),PO,
         2 P(30),BO(30),HO(30),MP,IFLAG,X(15),Y(15),REF(15,15),TRA(15,15)
         3 ,S(20,15),SO(15),DEP,DALT,IPRNT,NMAX,IO(20)
C
C
C             REF = REFLECTION COEFFICIENT
C             TRA = TRANSMISSION COEFFICIENT
C
          DO 1 I=1,N
          DO 1 J=1,N
```

```
      REF(I,J) = AL4*(X(I)*X(J) - Y(I)*Y(J))*RT(J) /(RT(I) + RT(J))
      IF(I.EQ.J) GO TO 2
      TRA(I,J) = AL4*(Y(I)*X(J) - X(I)*Y(J))*RT(J) /(RT(I) - RT(J))
      GO TO 1
    2 TRA(I,I) = 0.25*TR(I)/RT(I)
    1 CONTINUE
C
C          B AND H AT BOTTOM AND TOP
C
      DO 3 I=1,N
      BO(I)=0.0
      HO(I)=1.0/ RT(I)
      NI=N+I
      PO(NI)=C.0
      HO(NI)=0.0
      DO 4 J=1,N
      BO(NI) = BO(NI) + TRA(I,J)*WR(J)
    4 HO(NI) = HO(NI) + REF(I,J)*WR(J)
      BO(NI) = 2.0*BO(NI) + EXP(-T(2)/RT(I)) / RT(I)
    3 HO(NI) = 2.0*HO(NI)
C
      DO 5 I=1,N
      NI=N+I
      B(MP,I)=HO(NI)
      H(MP,I)=BO(NI)
      B(MP,NI)=HO(I)
    5 H(MP,NI)=BO(I)
      RETURN
      END
$IBFTC OUTPUT
      SUBROUTINE OUTPUT
C
      COMMON T(11119),N,MMAX,KMAX,K1MAX,RT(15),WT(15),DELTA,ALAM,THICK,
     1 NP,N2,NADD,AL2,AL4,WR(15),NEQ,TR(15),ALT,B(2C,30),H(2C,30),PO,
     2 P(30),BO(30),HO(30),MP,IFLAG,X(15),Y(15),REF(15,15),TRA(15,15)
     3 ,S(20,15),SO(15),DEP,DALT,IPRNT,NMAX,IO(20)
C
      WRITE(6,94)ALAM,THICK,(I,I=1,N)
C
    1 DEP=0.0
      M=1
      K=MP
      P(K)=0.0
      DO 4 I=1,N
    4 P(K)=P(K) + S(K,I)*WR(I)
      P(K)=2.0*P(K)
      WRITE(6,92)DEP, (B(K,I),I=1,N2)
      WRITE(6,91)      (H(K,I),I=1,N2)
      DEP=DEP + DALT
C
    5 DO 10 M=2,MP
      K=MP-M+1
C
      DO 3 I=1,N
      NI=N+I
```

```
      B(K,I)=H(M-1,NI)
    3 H(K,I)=B(M-1,NI)
C
    2 P(K)=0.0
      DO 12 I=1,N
   12 P(K)=P(K) + S(K,I)*WR(I)
      P(K)=2.0*P(K)
      WRITE(6,90)DEP,P(K),(B(K,I),I=1,N2)
      WRITE(6,91)          (H(K,I),I=1,N2)
C
   10 DEP=DEP + DALT
C
C             BOTTOM OF SLAB
      PO=0.0
      DO 11 I=1,N
   11 PO=PO + SO(I)*WR(I)
      PO=2.0*PO
      WRITE(6,90)DEP,PO,(BO(I),I=1,N2)
      WRITE(6,91)          (HO(I),I=1,N2)
   13 RETURN
C
   90 FORMAT(1H0,F8.2,1PE14.4,3H  B,1PE13.4,6E15.4/24X7E15.4)
   91 FORMAT(23X3H  H,1PE13.4,6E15.4/24X7E15.4)
   92 FORMAT(1H0,F8.2,6X8HINFINITE,3H  B,1PE13.4,6E15.4)/24X7E15.4)
   94 FORMAT(1H1 6X 8HALBEDO =,F6.4,5X11HTHICKNESS =,F8.4/
    1 1H0,8H   DEPTH,11X4HPHI  , 7I15)
      END
$ENTRY          MAIN
    7     19     10     10
25446046E-0112923441E-0029707742E-0050000000E 0070292258E 0087076559E 00
97455396E 00
64742484E-0113985269E-0019091502E-0020897958E-0019091502E-0013985269E-00
64742484E-01
        .005          1.0          .05
$IBSYS          ENDJOB
```

APPENDIX C

TABLES OF b, h AND Φ FUNCTIONS


We present computer produced tables of the b and
h functions for four different albedos. For each albedo,
we have selected several thicknesses ranging from small to
large. For each thickness, the optical depths considered
are equally spaced. The following is a summary of the cases
covered in the tables.


| Albedo | Thicknesses |
|--------|-------------|
| 0.2 | 0.5, 1.0 , |
| 0.5 | 1.0, 2.0, 3.0 |
| 0.9 | 1.0, 2.0, 3.0, 5.0 |
| 1.0 | 1.0, 2.0, 5.0, 10., 15., 20 . |


There are fifteen directions of propagation in the
slab for which intensities are tabulated. These are 90
degrees or horizontal, the seven angles measured from the
upward directed vertical: approximately 88.5, 82.6, 72.7,
60.0, 45.3, 29.5 and 13.0 degrees; and the same seven angles,
measured from the downward vertical.

The format of the tables is as follows: First the
albedo and thickness are given. Then, optical depths are
listed in the first column, and values of the Φ function,
i.e., b in the horizontal direction, in the second column,
corresponding to each depth. The last seven columns correspond to the seven angles 88.5, 82.6,...,13.0 degrees given

above, and are headed "1," "2," ..., "7." Consider, for
example, the column labeled "1." The four numbers given for
each depth are:  (1) the intensity  b,  due to isotropic
sources at the top, in an upward direction which is 88.5
degrees from the upward vertical; (2) the intensity  b  in
a downward direction which is 88.5 degrees from the downward
vertical; (3) the intensity  h, due to isotropic sources at
the bottom, in an upward direction which is 88.5 degrees
from the upward vertical; and (4) the intensity  h  in a
downward direction which is 88.5 degrees from the downward
vertical.  The numbers are printed in the  "E format" which
means that numbers are given with powers of ten.

   Let us illustrate the use of the tables.  We ask for
the intensity in a slab of thickness 0.5 with albedo 0.2,
at a depth 0.3 and in a direction which is 60 degrees from
the upward-directed vertical, due to isotropic sources on
unit net flux per unit horizontal area at the top.  From
page    , we obtain

$$b = 2.7305E\text{-}02 = 2.7305 \times 10^{-2} = .027305 \; .$$

   These tables are selections from extensive calculations
which consumed about one hour of computing time on the IBM
7044.

ALBEDO =0.2000          THICKNESS = 0.5000

| DEPTH | PHI | | 1 | 2 | 3 | 4 | 5 | 6 | 7 |
|---|---|---|---|---|---|---|---|---|---|
| 0. | INFINITE | B | 3.8005E-01 | 2.3023E-01 | 1.5206E-01 | 1.0777E-01 | 8.3313E-02 | 7.0106E-02 | 6.3838E-02 |
|  |  |  | 3.9299E 01 | 7.7379E 00 | 3.3661E 00 | 2.0000E 00 | 1.4226E 00 | 1.1484E 00 | 1.0261E 00 |
|  |  | H | 6.7721E-02 | 2.5065E-01 | 7.1943E-01 | 8.1161E-01 | 7.6602E-01 | 7.0586E-01 | 6.6911E-01 |
|  |  |  | 0. | 0. | 0. | 0. | 0. | 0. | 0. |
| 0.10 | 1.9439E-01 | B | 1.7669E-01 | 1.3374E-01 | 9.2093E-02 | 6.5538E-02 | 5.0630E-02 | 4.2264E-02 | 3.8737E-02 |
|  |  |  | 9.9466E-01 | 3.7117E 00 | 2.4857E 00 | 1.6898E 00 | 1.2725E 00 | 1.0555E 00 | 9.5446E-01 |
|  |  | H | 8.5879E-02 | 4.6037E-01 | 9.7880E-01 | 9.8097E-01 | 8.7218E-01 | 7.8304E-01 | 7.3368E-01 |
|  |  |  | 7.4840E-02 | 3.9261E-02 | 2.0655E-02 | 1.3065E-02 | 9.5471E-03 | 7.8068E-03 | 7.0157E-03 |
| 0.20 | 1.3659E-01 | B | 1.2716E-01 | 9.6491E-02 | 6.3486E-02 | 4.3975E-02 | 3.3503E-02 | 2.7965E-02 | 2.5367E-02 |
|  |  |  | 1.6420E-01 | 1.8002E 00 | 1.8211E 00 | 1.4127E 00 | 1.1251E 00 | 9.5839E-01 | 8.7711E-01 |
|  |  | H | 1.1117E-01 | 8.9317E-01 | 1.3342E 00 | 1.1780E 00 | 9.9161E-01 | 8.6726E-01 | 8.0312E-01 |
|  |  |  | 6.9924E-02 | 6.8048E-02 | 4.1012E-02 | 2.7305E-02 | 2.0417E-02 | 1.6883E-02 | 1.5249E-02 |
| 0.30 | 1.0322E-01 | B | 9.6924E-02 | 6.8048E-02 | 4.1012E-02 | 2.7305E-02 | 2.0417E-02 | 1.6883E-02 | 1.5249E-02 |
|  |  |  | 1.1117E-01 | 8.9317E-01 | 1.3342E 00 | 1.1780E 00 | 9.9161E-01 | 8.6726E-01 | 8.0312E-01 |
|  |  | H | 1.6420E-01 | 1.8002E 00 | 1.8211E 00 | 1.4127E 00 | 1.1251E 00 | 9.5839E-01 | 8.7711E-01 |
|  |  |  | 1.2716E-01 | 9.6491E-02 | 6.3486E-02 | 4.3975E-02 | 3.3503E-02 | 2.7965E-02 | 2.5367E-02 |
| 0.40 | 8.0572E-02 | B | 7.4840E-02 | 3.9261E-02 | 2.0655E-02 | 1.3065E-02 | 9.5471E-03 | 7.8068E-03 | 7.0157E-03 |
|  |  |  | 8.5879E-02 | 4.6037E-01 | 9.7880E-01 | 9.8097E-01 | 8.7218E-01 | 7.8304E-01 | 7.3368E-01 |
|  |  | H | 9.9464E-01 | 3.7117E 00 | 2.4857E 00 | 1.6898E 00 | 1.2725E 00 | 1.0554E 00 | 9.5444E-01 |
|  |  |  | 1.7669E-01 | 1.3374E-01 | 9.2093E-02 | 6.5538E-02 | 5.0630E-02 | 4.2264E-02 | 3.8737E-02 |
| 0.50 | 6.3559E-02 | B | 0. | 0. | 0. | 0. | 0. | 0. | 0. |
|  |  |  | 6.7721E-02 | 2.5045E-01 | 7.1943E-01 | 8.1161E-01 | 7.6602E-01 | 7.0586E-01 | 6.6911E-01 |
|  |  | H | 3.9299E 01 | 7.7379E 00 | 3.3661E 00 | 2.0000E 00 | 1.4226E 00 | 1.1484E 00 | 1.0261E 00 |
|  |  |  | 3.8005E-01 | 2.3023E-01 | 1.5206E-01 | 1.0777E-01 | 8.3313E-02 | 7.0106E-02 | 6.3838E-02 |

ALBEDO = -0.2000          THICKNESS = 1.0000

| DEPTH | PHI | | 1 | 2 | 3 | 4 | 5 | 6 | 7 |
|---|---|---|---|---|---|---|---|---|---|
| 0. | INFINITE | B | 3.8103E-01 | 2.3256E-01 | 1.6067E-01 | 1.1966E-01 | 9.5420E-02 | 8.2016E-02 | 7.5385E-02 |
| | | | 3.9299E 01 | 7.7379E 00 | 3.3661E 00 | 2.0000E 00 | 1.4226E 00 | 1.1484E 00 | 1.0261E 00 |
| | | H | 2.8196E-02 | 3.8523E-02 | 1.6536E-01 | 3.2627E-01 | 3.9763E-01 | 4.1621E-01 | 4.1786E-01 |
| | | | 0. | 0. | 0. | 0. | 0. | 0. | 0. |
| 0.10 | 1.9554E-01 | B | 1.7790E-01 | 1.3760E-01 | 1.0374E-01 | 7.9838E-02 | 6.4460E-02 | 5.5798E-02 | 5.1420E-02 |
| | | | 9.9572E-01 | 3.7183E 00 | 2.4800E 00 | 1.6900E 00 | 1.2726E 00 | 1.0555E 00 | 9.5454E-01 |
| | | H | 3.3764E-02 | 4.9445E-02 | 2.1978E-01 | 3.9199E-01 | 4.5392E-01 | 4.6328E-01 | 4.5983E-01 |
| | | | 3.0348E-02 | 1.6087E-02 | 8.4793E-03 | 5.3666E-03 | 3.9225E-03 | 3.2078E-03 | 2.8829E-03 |
| 0.20 | 1.3801E-01 | B | 1.2806E-01 | 1.0336E-01 | 7.9254E-02 | 6.1160E-02 | 4.9484E-02 | 4.2655E-02 | 3.9282E-02 |
| | | | 1.6555E-01 | 1.8011E 00 | 1.8217E 00 | 1.4131E 00 | 1.1254E 00 | 9.5863E-01 | 8.7732E-01 |
| | | H | 4.0253E-02 | 6.6474E-02 | 2.9370E-01 | 4.7099E-01 | 5.1793E-01 | 5.1537E-01 | 5.0572E-01 |
| | | | 3.6789E-02 | 2.6629E-02 | 1.6183E-02 | 1.0804E-02 | 8.0876E-03 | 6.9915E-03 | 6.0454E-03 |
| 0.30 | 1.0501E-01 | B | 9.8857E-02 | 8.1156E-02 | 6.2499E-02 | 4.7963E-02 | 3.8597E-02 | 3.3167E-02 | 3.0497E-02 |
| | | | 1.1286E-01 | 8.9448E-01 | 1.3351E 00 | 1.1786E 00 | 9.2070E-01 | 8.6764E-01 | 8.0347E-01 |
| | | H | 4.8152E-02 | 9.5551E-02 | 3.9449E-01 | 5.6598E-01 | 5.9070E-01 | 5.7298E-01 | 5.5583E-01 |
| | | | 4.3919E-02 | 3.5197E-02 | 2.3637E-02 | 1.6491E-02 | 1.2804E-02 | 1.0536E-02 | 9.5634E-03 |
| 0.40 | 8.2894E-02 | B | 7.8558E-02 | 6.5343E-02 | 4.9933E-02 | 3.7824E-02 | 3.0197E-02 | 2.5826E-02 | 2.3692E-02 |
| | | | 8.0050E-02 | 4.6209E-01 | 9.8001E-01 | 9.8183E-01 | 8.7285E-01 | 7.8361E-01 | 7.3419E-01 |
| | | H | 5.8038E-02 | 1.4400E-01 | 5.3231E-01 | 6.8017E-01 | 6.7332E-01 | 6.3659E-01 | 6.1045E-01 |
| | | | 5.2703E-02 | 4.3698E-02 | 3.1156E-02 | 2.2663E-02 | 1.7628E-02 | 1.4869E-02 | 1.3552E-02 |
| 0.50 | 6.7086E-02 | B | 6.3861E-02 | 5.3366E-02 | 3.9890E-02 | 2.9627E-02 | 2.3383E-02 | 1.9873E-02 | 1.8178E-02 |
| | | | 7.0818E-02 | 2.5279E-01 | 7.2110E-01 | 8.1733E-01 | 7.6697E-01 | 7.0667E-01 | 6.6985E-01 |
| | | H | 7.0818E-02 | 2.5279E-01 | 7.2110E-01 | 8.1733E-01 | 7.6697E-01 | 7.0667E-01 | 6.9085E-01 |
| | | | 6.3861E-02 | 5.3366E-02 | 3.9890E-02 | 2.9627E-02 | 2.3383E-02 | 1.9873E-02 | 1.8178E-02 |
| 0.60 | 5.5203E-02 | B | 5.2703E-02 | 4.3698E-02 | 3.1156E-02 | 2.2663E-02 | 1.7628E-02 | 1.4869E-02 | 1.3552E-02 |
| | | | 5.8038E-02 | 1.4400E-01 | 5.3231E-01 | 6.8017E-01 | 6.7332E-01 | 6.3659E-01 | 6.1042E-01 |
| | | H | 8.0050E-02 | 4.6209E-01 | 9.8001E-01 | 9.8183E-01 | 8.7285E-01 | 7.8361E-01 | 7.3419E-01 |
| | | | 7.8558E-02 | 6.5343E-02 | 4.9933E-02 | 3.7824E-02 | 3.0197E-02 | 2.5826E-02 | 2.3692E-02 |
| 0.70 | 4.5919E-02 | B | 4.3919E-02 | 3.5197E-02 | 2.3637E-02 | 1.6491E-02 | 1.2804E-02 | 1.0536E-02 | 9.5634E-03 |
| | | | 4.8152E-02 | 9.5551E-02 | 3.9449E-01 | 5.6598E-01 | 5.9070E-01 | 5.7298E-01 | 5.5583E-01 |
| | | H | 1.1286E-01 | 8.9448E-01 | 1.3351E 00 | 1.1786E 00 | 9.2070E-01 | 8.6764E-01 | 8.0347E-01 |
| | | | 9.8857E-02 | 8.1156E-02 | 6.2499E-02 | 4.7963E-02 | 3.8597E-02 | 3.3167E-02 | 3.0497E-02 |
| 0.80 | 3.8643E-02 | B | 3.6789E-02 | 2.6629E-02 | 1.6183E-02 | 1.0804E-02 | 8.0876E-03 | 6.9915E-03 | 6.0454E-03 |
| | | | 4.0253E-02 | 6.6474E-02 | 2.9370E-01 | 4.7099E-01 | 5.1793E-01 | 5.1537E-01 | 5.0572E-01 |
| | | H | 1.6555E-01 | 1.8011E 00 | 1.8217E 00 | 1.4131E 00 | 1.1254E 00 | 9.5863E-01 | 8.7732E-01 |
| | | | 1.2806E-01 | 1.0336E-01 | 7.9254E-02 | 6.1160E-02 | 4.9484E-02 | 4.2655E-02 | 3.9282E-02 |
| 0.90 | 3.2254E-02 | B | 3.0348E-02 | 1.6087E-02 | 8.4793E-03 | 5.3666E-03 | 3.9225E-03 | 3.2078E-03 | 2.8829E-03 |
| | | | 3.3764E-02 | 4.9445E-02 | 2.1978E-01 | 3.9199E-01 | 4.5392E-01 | 4.6328E-01 | 4.5983E-01 |
| | | H | 9.9272E-01 | 3.7183E 00 | 2.4800E 00 | 1.6900E 00 | 1.2726E 00 | 1.0555E 00 | 9.5454E-01 |
| | | | 1.7790E-01 | 1.3760E-01 | 1.0374E-01 | 7.9838E-02 | 6.4460E-02 | 5.5798E-02 | 5.1420E-02 |
| 1.00 | 2.6814E-02 | B | 0. | 0. | 0. | 0. | 0. | 0. | 0. |
| | | | 2.8196E-02 | 3.8523E-02 | 1.6536E-01 | 3.2627E-01 | 3.9763E-01 | 4.1621E-01 | 4.1786E-01 |
| | | H | 3.9299E 01 | 7.7379E 00 | 3.3661E 00 | 2.0000E 00 | 1.4226E 00 | 1.1484E 00 | 1.0261E 00 |
| | | | 3.8103E-01 | 2.3256E-01 | 1.6067E-01 | 1.1966E-01 | 9.5420E-02 | 8.2016E-02 | 7.5385E-02 |

ALBEDO =0.5000     THICKNESS = 1.0000

| DEPTH | PHI | | 1 | 2 | 3 | 4 | 5 | 6 | 7 |
|---|---|---|---|---|---|---|---|---|---|
| 0. | INFINITE | R | 1.0905 00 | 6.6012E-01 | 4.7250E-01 | 3.5916E-01 | 2.9005E-01 | 2.5016E-01 | 2.3052E-01 |
| | | H | 1.9299E-01 | 7.3370E 00 | 3.3631E 00 | 2.7000E 00 | 1.4284E 00 | 1.1484E 00 | 1.0251E 00 |
| | | | 1.0319E-01 | 1.2954E-01 | 2.8427E-01 | 4.5445E-01 | 5.2035E-01 | 5.3216E-01 | 5.2931E-01 |
| | | | 0. | 0. | 0. | 0. | 0. | 0. | 0. |
| 0.10 | 5.7125E-01 | R | 5.2671E-01 | 4.2143E-01 | 3.3250E 00 | 2.5511E-01 | 2.0818E-01 | 1.8035E-01 | 1.6646E-01 |
| | | H | 1.4121E 00 | 3.9855E 00 | 2.6319E 00 | 1.7834E 00 | 1.3412E 00 | 1.1117E 00 | 1.0051E 00 |
| | | | 1.2490E-01 | 1.5777E-01 | 3.5555E-01 | 5.3151E-01 | 5.8417E-01 | 5.8393E-01 | 5.7486E-01 |
| | | | 1.1024E-01 | 5.8252E-02 | 3.0084E-02 | 1.7416E-02 | 1.4190E-02 | 1.1604E-02 | 1.0429E-02 |
| 0.20 | 4.2479E-01 | R | 4.0927E-01 | 3.3117E-01 | 2.5979E-01 | 2.0268E-01 | 1.5491E-01 | 1.4254E-01 | 1.3144E-01 |
| | | H | 4.7112E-01 | 2.0972E 00 | 2.0142E 00 | 1.5444E 00 | 1.2805E 00 | 1.0441E 00 | 9.5472E-01 |
| | | | 1.4112E-01 | 1.9395E-01 | 4.2675E-01 | 6.2987E-01 | 6.3591E-01 | 6.3941E-01 | 6.2291E-01 |
| | | | 1.3735E-01 | 9.6715E-02 | 5.8752E-02 | 3.927E-02 | 2.8934E-02 | 2.2866E-02 | 2.1941E-02 |
| 0.30 | 3.3736E-01 | R | 3.5492E-01 | 2.6974E-01 | 2.1137E-01 | 1.6351E-01 | 1.2214E-01 | 1.2376E-01 | 1.0469E-01 |
| | | H | 3.5707E-01 | 1.1661E 00 | 1.5499E 00 | 1.3605E 00 | 1.1125E 00 | 9.7173E-01 | 9.0942E-01 |
| | | | 1.7140E-01 | 2.4494E-01 | 5.6077E-01 | 7.2446E-01 | 7.3071E-01 | 6.9874E-01 | 6.7134E-01 |
| | | | 1.5595E-01 | 1.2757E-01 | 8.5752E-02 | 5.7235E-02 | 4.5622E-02 | 3.8147E-02 | 3.4620E-02 |
| 0.40 | 2.7612E-01 | R | 2.6370E-01 | 2.2326E-01 | 1.7325E-01 | 1.3194E-01 | 1.0559E-01 | 9.0407E-02 | 8.2992E-02 |
| | | H | 2.9052E-01 | 7.0109E-01 | 1.1892E 00 | 1.1596E 00 | 1.0077E 00 | 8.4934E-01 | 8.4051E-01 |
| | | | 2.0308E-01 | 3.2391E-01 | 7.0478E-01 | 7.7466E-01 | 8.1591E-01 | 7.2066E-01 | 7.2744E-01 |
| | | | 1.6633E-01 | 1.3617E-01 | 1.1231E-01 | 5.1215E-02 | 8.3375E-02 | 5.3622E-02 | 4.8711E-02 |
| 0.50 | 2.3033E-01 | R | 2.2665E-01 | 1.8730E-01 | 1.0113E-01 | 1.0945E-01 | 8.3807E-02 | 7.0661E-02 | 5.4651E-02 |
| | | H | 2.4145E-01 | 4.5730E-01 | 8.2705E-01 | 8.6792E-01 | 9.0793E-01 | 8.2990E-01 | 7.8310E-01 |
| | | | 2.4286E-01 | 4.5770E-01 | 2.5705E-01 | 9.5529E-01 | 8.2074E-01 | 8.2905E-01 | 7.7310E-01 |
| | | | 2.2645E-01 | 1.8730E-01 | 1.4115E-01 | 1.0314E-01 | 9.3007E-02 | 7.2061E-02 | 6.5551E-02 |
| 0.60 | 1.8924E-01 | R | 1.8639E-01 | 1.5617E-01 | 1.1345E-01 | 8.1271E-02 | 6.3367E-02 | 5.7462E-02 | 4.2945E-02 |
| | | H | 2.0277E-01 | 3.2591E-01 | 7.3890E-01 | 9.4511E-01 | 8.5111E-01 | 7.6604E-01 | 7.2744E-01 |
| | | | 2.2405E-01 | 7.01098E-01 | 1.5275E-01 | 1.1414E 00 | 1.0077E 00 | 8.5935E-01 | 7.4095E-01 |
| | | | 2.4670E-01 | 2.2380E-01 | 1.7325E-01 | 1.5194E 00 | 1.0595E 00 | 9.0407E-02 | 6.2932E-02 |
| 0.70 | 1.6648E-01 | R | 1.5895E-01 | 1.2727E-01 | 8.5555E-02 | 5.9722E-02 | 4.5632E-02 | 3.8173E-02 | 3.4620E-02 |
| | | H | 1.7128E-01 | 2.4646E-01 | 5.6577E-01 | 7.2886E-01 | 7.2071E-01 | 3.9787E-01 | 8.9247E-01 |
| | | | 3.2490E-01 | 1.1681E 00 | 1.4680E 00 | 1.3580E 00 | 1.1125E 00 | 9.7173E-01 | 8.9942E-01 |
| | | | 3.2490E-01 | 2.6974E-01 | 2.1137E 00 | 1.6351E-01 | 1.3246E 00 | 1.1376E-01 | 1.0469E-01 |
| 0.80 | 1.3395E-01 | R | 1.3174E-01 | 1.1272E-01 | 5.8735E-02 | 3.0717E-02 | 2.3554E-02 | 2.4293E-02 | 2.1941E-02 |
| | | H | 1.4572E-01 | 1.9386E-01 | 4.4675E-01 | 6.2607E-01 | 6.5391E-01 | 6.3911E-01 | 6.2291E-01 |
| | | | 4.0027E-01 | 2.0972E 00 | 2.0142E 00 | 1.5484E 00 | 1.2250E 00 | 1.0441E 00 | 9.5472E-01 |
| | | | 5.2471E-01 | 3.3117E-01 | 2.5579E 00 | 2.0268E-01 | 1.6642E-01 | 1.4294E-01 | 1.3144E-01 |
| 0.90 | 1.1742E-01 | R | 1.1623E-01 | 5.8232E-02 | 3.0564E-02 | 1.7418E-02 | 1.4190E-02 | 1.1604E-02 | 1.0429E-02 |
| | | H | 1.2290E-01 | 1.5777E-01 | 3.5555E-01 | 5.3151E-01 | 5.8417E-01 | 5.8393E-01 | 5.7466E-01 |
| | | | 1.4121E 00 | 3.9855E 00 | 2.6319E 00 | 1.7834E 00 | 1.3412E 00 | 1.1117E 00 | 1.0051E 00 |
| | | | 5.2671E-01 | 4.2143E-01 | 3.2659E-01 | 2.5511E-01 | 2.0818E-01 | 1.8035E-01 | 1.6646E-01 |
| 1.00 | 9.5507E-02 | R | 0. | 1.2954E-01 | 2.8427E-01 | 4.5445E-01 | 5.2035E-01 | 5.3216E-01 | 5.2931E-01 |
| | | H | 3.9299E-01 | 7.3370E 00 | 3.3631E 00 | 2.0000E 00 | 1.4284E 00 | 1.1484E 00 | 1.0261E 00 |
| | | | 1.0400E 00 | 6.6012E-01 | 4.7250E-01 | 3.5916E-01 | 2.9005E-01 | 2.5016E-01 | 2.3052E-01 |

ALBEDO = 0.5000   THICKNESS = 2.0000

| DEPTH | PHI | | 1 | 2 | 3 | 4 | 5 | 6 | 7 |
|---|---|---|---|---|---|---|---|---|---|
| 0. | INFINITE | B | 1.0336E 00 | 6.6656E-01 | 4.8100E-01 | 3.7414E-01 | 3.0922E-01 | 2.7108E-01 | 2.5200E-01 |
| | | H | 3.9299E 01 | 7.7379E 00 | 5.3661E 00 | 2.0000E 00 | 1.4226E 00 | 1.1484E 00 | 1.0261E 00 |
| | | | 3.0877E-02 | 3.7160E-02 | 5.4915E-02 | 1.0818E-01 | 1.6978E-01 | 2.1016E-01 | 2.2922E-01 |
| | | | 0. | 0. | 0. | 0. | 0. | 0. | 0. |
| 0.10 | 5.7539E-01 | B | 5.3102E-01 | 4.2673E-01 | 3.3686E-01 | 2.7258E-01 | 2.2970E-01 | 2.0335E-01 | 1.8987E-01 |
| | | H | 1.4160E 00 | 3.2875E 00 | 2.6330E 00 | 1.7841E 00 | 1.3417E 00 | 1.1121E 00 | 1.0055E 00 |
| | | | 3.6684E-02 | 4.3215E-02 | 6.4011E-02 | 1.2499E-01 | 1.9080E-01 | 2.3181E-01 | 2.5050E-01 |
| | | | 3.3242E-02 | 1.7634E-02 | 9.2946E-03 | 5.8827E-03 | 4.2997E-03 | 3.5163E-03 | 3.1602E-03 |
| 0.20 | 4.2966E-01 | B | 4.0533E-01 | 3.3741E-01 | 2.7236E-01 | 2.2303E-01 | 1.8904E-01 | 1.6781E-01 | 1.5688E-01 |
| | | H | 4.7640E-01 | 2.1006E 00 | 2.0205E 00 | 1.5498E 00 | 1.2291E 00 | 1.0449E 00 | 9.5549E-01 |
| | | | 4.2497E-02 | 4.9636E-02 | 7.4451E-02 | 1.4426E-01 | 2.1416E-01 | 2.5539E-01 | 2.7346E-01 |
| | | | 3.9482E-02 | 2.8847E-02 | 1.7569E-02 | 1.1737E-02 | 8.7886E-03 | 7.2725E-03 | 6.5707E-03 |
| 0.30 | 3.4301E-01 | B | 3.2639E-01 | 2.7717E-01 | 2.2694E-01 | 1.8719E-01 | 1.5916E-01 | 1.4147E-01 | 1.3231E-01 |
| | | H | 3.6333E-01 | 1.1725E 00 | 1.5519E 00 | 1.3381E 00 | 1.1168E 00 | 9.7307E-01 | 8.9963E-01 |
| | | | 4.8751E-02 | 5.6688E-02 | 8.6710E-02 | 1.6648E-01 | 2.4018E-01 | 2.8113E-01 | 2.9826E-01 |
| | | | 4.5499E-02 | 3.7215E-02 | 2.5168E-02 | 1.7601E-02 | 1.3466E-02 | 1.1262E-02 | 1.0225E-02 |
| 0.40 | 2.8272E-01 | B | 2.7055E-01 | 2.3288E-01 | 1.9253E-01 | 1.5951E-01 | 1.3580E-01 | 1.2074E-01 | 1.1292E-01 |
| | | H | 2.9690E-01 | 7.0660E-01 | 1.1968E 00 | 1.1518E 00 | 1.0099E 00 | 9.0120E-01 | 8.4220E-01 |
| | | | 5.5624E-02 | 6.4553E-02 | 1.0132E-01 | 1.9221E-01 | 2.6921E-01 | 3.0921E-01 | 3.2506E-01 |
| | | | 5.2038E-02 | 4.4538E-02 | 3.2428E-02 | 2.3564E-02 | 1.8372E-02 | 1.5513E-02 | 1.4146E-02 |
| 0.50 | 2.3799E-01 | B | 2.4863E-01 | 1.9865E-01 | 1.6529E-01 | 1.3724E-01 | 1.1683E-01 | 1.0382E-01 | 9.7061E-02 |
| | | H | 2.4805E-01 | 4.6405E-01 | 9.2852E-01 | 9.8989E-01 | 9.1028E-01 | 8.3152E-01 | 7.8532E-01 |
| | | | 6.3276E-02 | 7.3424E-02 | 1.1897E-01 | 2.2207E-01 | 3.0160E-01 | 3.3986E-01 | 3.5398E-01 |
| | | | 5.9273E-02 | 5.1740E-02 | 3.9634E-02 | 2.9725E-02 | 2.3563E-02 | 2.0069E-02 | 1.8374E-02 |
| 0.60 | 2.0321E-01 | B | 1.9572E-01 | 1.7114E-01 | 1.4304E-01 | 1.1890E-01 | 1.0101E-01 | 8.9644E-02 | 8.3752E-02 |
| | | H | 2.1157E-01 | 3.3134E-01 | 7.2572E-01 | 8.5022E-01 | 8.1868E-01 | 7.6513E-01 | 7.3016E-01 |
| | | | 7.1875E-02 | 8.3531E-02 | 1.4055E-01 | 2.5681E-01 | 3.3774E-01 | 3.7327E-01 | 3.8519E-01 |
| | | | 6.7367E-02 | 5.9333E-02 | 4.7037E-02 | 3.6199E-02 | 2.9111E-02 | 2.4985E-02 | 2.2957E-02 |
| 0.70 | 1.7520E-01 | B | 1.5905E-01 | 1.4851E-01 | 1.2446E-01 | 1.0321E-01 | 8.7454E-02 | 7.7554E-02 | 7.2382E-02 |
| | | H | 1.8198E-01 | 2.5355E-01 | 5.7792E-01 | 7.3023E-01 | 7.3510E-01 | 7.0256E-01 | 6.7733E-01 |
| | | | 8.1612E-02 | 9.5175E-02 | 1.6722E-01 | 2.9729E-01 | 3.7806E-01 | 4.0970E-01 | 4.1880E-01 |
| | | | 7.6499E-02 | 6.7646E-02 | 5.4855E-02 | 4.3109E-02 | 3.5100E-02 | 3.0330E-02 | 2.7959E-02 |
| 0.80 | 1.5208E-01 | B | 1.1694E-01 | 1.2953E-01 | 1.0847E-01 | 8.9827E-02 | 7.5925E-02 | 6.7093E-02 | 6.2537E-02 |
| | | H | 1.5771E-01 | 2.0413E-01 | 4.5601E-01 | 6.2739E-01 | 6.5924E-01 | 6.4403E-01 | 6.2718E-01 |
| | | | 9.2715E-02 | 1.0878E-01 | 2.0056E-01 | 3.4450E-01 | 4.2300E-01 | 4.4932E-01 | 4.5497E-01 |
| | | | 8.6875E-02 | 7.6941E-02 | 5.3339E-02 | 5.0592E-02 | 4.1629E-02 | 3.6187E-02 | 3.3454E-02 |
| 0.90 | 1.3268E-01 | B | 1.2831E-01 | 1.1338E-01 | 9.5917E-02 | 7.8180E-02 | 6.5771E-02 | 5.7941E-02 | 5.3921E-02 |
| | | H | 1.3742E-01 | 1.7005E-01 | 3.6542E-01 | 5.3334E-01 | 5.9063E-01 | 5.8356E-01 | 5.7586E-01 |
| | | | 1.0746E-01 | 1.2497E-01 | 2.4262E-01 | 4.6405E-01 | 4.7306E-01 | 4.9237E-01 | 4.9380E-01 |
| | | | 9.3747E-02 | 8.7488E-02 | 7.2685E-02 | 5.8605E-02 | 4.8818E-02 | 4.2656E-02 | 3.9594E-02 |
| 1.00 | 1.1817E-01 | B | 1.1243E-01 | 9.9503E-02 | 8.3130E-02 | 6.7292E-02 | 5.6808E-02 | 4.9858E-02 | 4.6310E-02 |
| | | H | 1.2021E-01 | 1.4479E-01 | 2.9635E-01 | 4.5405E-01 | 5.2976E-01 | 5.3905E-01 | 5.3541E-01 |
| | | | 1.2021E-01 | 1.4479E-01 | 2.9635E-01 | 4.6405E-01 | 5.2876E-01 | 5.3905E-01 | 5.3541E-01 |
| | | | 1.1243E-01 | 3.9503E-01 | 9.4105E-02 | 6.7292E-02 | 5.6808E-02 | 4.9858E-02 | 4.6310E-02 |
| 1.10 | 1.0198E-01 | B | 9.8747E-02 | 8.7468E-02 | 7.2585E-02 | 5.8805E-02 | 4.8818E-02 | 4.2656E-02 | 3.9534E-02 |
| | | H | 1.0546E-01 | 1.2497E-01 | 6.4266E-01 | 3.9064E-01 | 4.7306E-01 | 4.9237E-01 | 4.9380E-01 |

| μ | (param) | | C1 | C2 | C3 | C4 | C5 | C6 | C7 |
|---|---------|---|----|----|----|----|----|----|----|
| 1.20 | 9.9694E-02 | β | 1.3742E-01 | 1.7005E-01 | 3.6542E-02 | 5.3736E-01 | 5.9063E-01 | 5.8956E-01 | 5.7986E-01 |
|  |  | x | 1.2831E-01 | 1.1338E-01 | 9.5017E-02 | 7.8180E-02 | 6.5771E-02 | 5.7941E-02 | 5.3921E-02 |
|  |  | β | 8.6875E-02 | 7.6941E-02 | 6.1339E-02 | 5.0592E-02 | 4.1629E-02 | 3.6187E-02 | 3.3454E-02 |
|  |  | x | 9.2715E-02 | 1.0873E-01 | 2.0050E-01 | 3.4430E-01 | 4.2300E-01 | 4.4932E-01 | 4.5497E-01 |
| 1.30 | 7.8971E-02 | β | 1.5771E-01 | 2.0413E-01 | 4.5501E-01 | 6.2739E-01 | 6.5924E-01 | 6.4403E-01 | 6.2718E-01 |
|  |  | x | 1.4694E-01 | 1.2953E-01 | 1.0863E-01 | 9.8927E-02 | 7.5925E-02 | 6.7093E-02 | 6.2537E-02 |
|  |  | β | 7.6499E-02 | 6.7606E-02 | 5.4865E-02 | 4.3100E-02 | 3.5100E-02 | 3.0330E-02 | 2.7959E-02 |
|  |  | x | 8.1612E-02 | 9.5175E-02 | 1.6722E-01 | 2.9709E-01 | 3.7800E-01 | 4.0970E-01 | 4.1880E-01 |
| 1.40 | 6.9551E-02 | β | 1.8198E-01 | 2.5355E-01 | 5.7202E-01 | 7.3023E-01 | 7.3510E-01 | 7.0250E-01 | 6.7733E-01 |
|  |  | x | 1.6905E-01 | 1.4851E-01 | 1.2444E-01 | 1.0321E-01 | 8.7546E-02 | 7.7544E-02 | 7.2382E-02 |
|  |  | β | 6.7367E-02 | 5.9333E-02 | 4.7037E-02 | 3.6199E-02 | 2.9111E-02 | 2.4985E-02 | 2.2957E-02 |
|  |  | x | 7.1875E-02 | 8.3511E-02 | 1.4055E-01 | 2.5681E-01 | 3.3744E-01 | 3.7329E-01 | 3.8519E-01 |
| 1.50 | 6.1216E-02 | β | 5.9273E-02 | 5.1740E-02 | 3.9636E-02 | 2.9725E-02 | 2.3563E-02 | 2.0069E-02 | 1.8374E-02 |
|  |  | x | 6.1276E-02 | 7.3426E-02 | 1.1897E-01 | 2.2207E-01 | 3.0105E-01 | 3.1986E-01 | 7.8532E-01 |
|  |  | β | 2.4855E-01 | 4.6405E-01 | 9.2852E-01 | 9.8989E-01 | 9.1028E-01 | 8.3152E-01 | 7.8532E-01 |
|  |  | x | 2.2863E-01 | 1.9863E-01 | 1.6529E-01 | 1.3724E-01 | 1.1683E-01 | 1.0382E-01 | 9.7061E-02 |
| 1.60 | 5.3783E-02 | β | 5.2038E-02 | 4.4530E-02 | 3.2428E-02 | 2.3564E-02 | 1.8372E-02 | 1.5513E-02 | 1.4146E-02 |
|  |  | x | 5.5624E-02 | 6.4553E-02 | 1.0132E-01 | 1.9221E-01 | 2.6921E-01 | 3.0921E-01 | 3.2506E-01 |
|  |  | β | 2.9690E-01 | 7.0640E-02 | 1.1968E-01 | 1.1518E-00 | 1.0099E-00 | 9.0120E-01 | 8.4220E-01 |
|  |  | x | 2.7055E-01 | 2.3288E-01 | 1.0253E-01 | 1.5951E-01 | 1.3580E-01 | 1.2074E-01 | 1.1292E-01 |
| 1.70 | 4.7098E-02 | β | 4.5449E-02 | 3.7215E-02 | 2.5168E-02 | 1.7601E-02 | 1.3466E-02 | 1.1262E-02 | 1.0225E-02 |
|  |  | x | 4.8751E-02 | 5.6688E-02 | 8.6710E-02 | 1.6644E-01 | 2.4018E-01 | 2.8113E-01 | 2.9926E-01 |
|  |  | β | 3.6333E-01 | 1.1725E-00 | 1.5519E-00 | 1.3381E-00 | 1.1168E-00 | 9.7307E-01 | 6.9953E-01 |
|  |  | x | 3.2639E-01 | 2.7717E-01 | 2.2694E-01 | 1.8719E-01 | 1.5916E-01 | 1.4147E-01 | 1.3231E-01 |
| 1.80 | 4.0970E-02 | β | 3.9442E-02 | 2.8347E-02 | 1.7539E-02 | 1.1737E-02 | 8.7886E-03 | 7.2725E-03 | 6.5707E-03 |
|  |  | x | 4.2497E-02 | 4.9636E-02 | 7.4451E-02 | 1.4426E-01 | 2.1416E-01 | 2.5339E-01 | 2.7346E-01 |
|  |  | β | 4.7640E-01 | 2.1000E-00 | 2.0205E-00 | 1.5498E-00 | 1.2291E-00 | 1.0449E-00 | 9.5549E-01 |
|  |  | x | 4.0533E-01 | 3.3741E-01 | 2.7236E-01 | 2.2303E-01 | 1.8904E-01 | 1.6781E-01 | 1.5688E-01 |
| 1.90 | 3.5241E-02 | β | 3.3252E-02 | 1.7634E-02 | 6.2946E-03 | 5.8827E-03 | 4.2397E-03 | 5.5163E-03 | 3.1602E-03 |
|  |  | x | 3.6644E-02 | 4.3215E-02 | 6.4018E-02 | 1.2499E-01 | 1.9080E-01 | 2.3181E-01 | 2.5050E-01 |
|  |  | β | 1.4160E-00 | 3.9875E-00 | 2.6330E-00 | 1.7841E-00 | 1.3417E-00 | 1.1121E-00 | 1.0055E-00 |
|  |  | x | 5.3102E-01 | 4.2673E-01 | 3.3686E-01 | 2.7258E-01 | 2.2970E-01 | 2.0335E-01 | 1.8987E-01 |
| 2.00 | 2.9233E-02 | β | 3.0877E-02 | 3.1160E-02 | 5.4315E-02 | 1.0318E-01 | 1.6978E-01 | 2.1010E-01 | 2.2922E-01 |
|  |  | x | 3.9294E-01 | 7.7379E-00 | 3.3661E-00 | 2.0000E-00 | 1.4226E-00 | 1.1486E-00 | 1.0261E-00 |
|  |  | β | 1.0336E-00 | 6.6456E-00 | 4.8100E-01 | 3.7414E-01 | 3.0922E-01 | 2.7100E-01 | 2.5209E-01 |

ALBEDO =0.5000    THICKNESS = 3.0000

| DEPTH | PHI | | 1 | 2 | 3 | 4 | 5 | 6 | 7 |
|---|---|---|---|---|---|---|---|---|---|
| 0. | INFINITE | B | 1.0360E 00 | 6.6500E-01 | 4.8163E-01 | 3.7534E-01 | 3.1122E-01 | 2.7370E-01 | 2.5494E-01 |
| | | | 3.9299E 01 | 7.7379E 00 | 3.3661E 00 | 2.0000E 00 | 1.4226E 00 | 1.1484E 00 | 1.0261E 00 |
| | | H | 1.0521E-02 | 1.2546E-02 | 1.6889E-02 | 3.0025E-02 | 5.5464E-02 | 8.0255E-02 | 9.5176E-02 |
| | | | 0. | 0. | 0. | 0. | 0. | 0. | 0. |
| 0.50 | 2.3871E-01 | B | 2.2937E-01 | 1.9948E-01 | 1.6656E-01 | 1.3963E-01 | 1.2038E-01 | 1.0806E-01 | 1.0162E-01 |
| | | | 2.4934E-01 | 4.6466E-01 | 9.2899E-01 | 9.9024E-01 | 9.1056E-01 | 8.3175E-01 | 7.8554E-01 |
| | | H | 2.0760E-02 | 2.3704E-02 | 3.1477E-02 | 5.7438E-02 | 9.8139E-02 | 1.3133E-01 | 1.4929E-01 |
| | | | 1.9540E-02 | 1.7171E-02 | 1.3209E-07 | 9.9240E-03 | 7.8729E-03 | 6.7080E-03 | 6.1426E-03 |
| 1.00 | 1.1749E-01 | B | 1.1379E-01 | 1.0110E-01 | 8.5971E-02 | 7.2830E-02 | 6.3044E-02 | 5.6654E-02 | 5.3285E-02 |
| | | | 1.2149E-01 | 1.4593E-01 | 2.9731E-01 | 4.6484E-01 | 5.2942E-01 | 5.3963E-01 | 5.3595E-01 |
| | | H | 3.6906E-02 | 4.2071E-02 | 5.8867E-02 | 1.1142E-01 | 1.7253E-01 | 2.1261E-01 | 2.3152E-01 |
| | | | 3.4827E-02 | 3.1220E-02 | 2.6400E-02 | 2.1716E-02 | 1.8225E-02 | 1.6024E-02 | 1.4897E-02 |
| 1.50 | 6.3758E-02 | B | 6.1910E-02 | 5.5422E-02 | 4.7379E-02 | 4.0027E-02 | 3.4413E-02 | 3.0742E-02 | 2.8814E-02 |
| | | | 6.5731E-02 | 7.5587E-02 | 1.2080E-01 | 2.2360E-01 | 3.0291E-01 | 3.4102E-01 | 3.5507E-01 |
| | | H | 6.5731E-02 | 7.5587E-02 | 1.2080E-01 | 2.2360E-01 | 3.0291E-01 | 3.4102E-01 | 3.5507E-01 |
| | | | 6.1910E-02 | 5.5422E-02 | 4.7379E-02 | 4.0027E-02 | 3.4413E-02 | 3.0742E-02 | 2.8814E-02 |
| 2.00 | 3.5836E-02 | B | 3.4827E-02 | 3.1220E-02 | 2.6400E-02 | 2.1716E-02 | 1.8225E-02 | 1.6024E-02 | 1.4897E-02 |
| | | | 3.6906E-02 | 4.2071E-02 | 5.8867E-02 | 1.1142E-01 | 1.7253E-01 | 2.1261E-01 | 2.3152E-01 |
| | | H | 1.2149E-01 | 1.4593E-01 | 2.9731E-01 | 4.6484E-01 | 5.2942E-01 | 5.3963E-01 | 5.3595E-01 |
| | | | 1.1379E-01 | 1.0110E-01 | 8.5971E-02 | 7.2830E-02 | 6.3044E-02 | 5.6654E-02 | 5.3285E-02 |
| 2.50 | 2.0135E-02 | B | 1.9540E-02 | 1.7171E-02 | 1.3209E-02 | 9.9240E-03 | 7.8729E-03 | 6.7080E-03 | 6.1426E-03 |
| | | | 2.0760E-02 | 2.3704E-02 | 3.1477E-02 | 5.7438E-02 | 9.8139E-02 | 1.3133E-01 | 1.4929E-01 |
| | | H | 2.4934E-01 | 4.6466E-01 | 9.2899E-01 | 9.9024E-01 | 9.1056E-01 | 8.3175E-01 | 7.8554E-01 |
| | | | 2.2937E-01 | 1.9948E-01 | 1.6656E-01 | 1.3963E-01 | 1.2038E-01 | 1.0806E-01 | 1.0162E-01 |
| 3.00 | 9.9773E-03 | B | 0. | 0. | 0. | 0. | 0. | 0. | 0. |
| | | | 1.0521E-02 | 1.2546E-02 | 1.6889E-02 | 3.0025E-02 | 5.5464E-02 | 8.0255E-02 | 9.5176E-02 |
| | | H | 3.9299E 01 | 7.7379E 00 | 3.3661E 00 | 2.0000E 00 | 1.4226E 00 | 1.1484E 00 | 1.0261E 00 |
| | | | 1.0360E 00 | 6.6500E-01 | 4.8163E-01 | 3.7534E-01 | 3.1122E-01 | 2.7370E-01 | 2.5494E-01 |

ALBEDO = 0.9000    THICKNESS = 1.0000

| DEPTH | PHI | | 1 | 2 | 3 | 4 | 5 | 6 | 7 |
|---|---|---|---|---|---|---|---|---|---|
| 0. | INFINITE | A | 2.1553E 00 | 1.5137E 00 | 1.1600E 00 | 9.1559E-01 | 7.5399E-01 | 6.5685E-01 | 6.0809E-01 |
| | | | 3.2249E 01 | 7.7379E 00 | 3.3651E 00 | 2.0000E 00 | 1.4226E 00 | 1.1484E 00 | 1.0261E 00 |
| | | H | 3.4694E-01 | 4.3283E-01 | 6.4485E-01 | 8.2369E-01 | 8.6478E-01 | 8.5082E-01 | 8.3242E-01 |
| | | | 0. | 0. | 0. | 0. | 0. | 0. | 0. |
| 0.10 | 1.3652E 00 | A | 1.2885E 00 | 1.0970E 00 | 8.9439E-01 | 7.1770E-01 | 5.9356E-01 | 5.1767E-01 | 4.7938E-01 |
| | | | 2.2481E 00 | 4.4915E 00 | 2.9055E 00 | 1.9580E 00 | 1.4692E 00 | 1.2166E 00 | 1.0994E 00 |
| | | H | 4.2701E-01 | 5.1472E-01 | 7.6215E-01 | 9.2487E-01 | 9.4086E-01 | 9.0966E-01 | 8.8267E-01 |
| | | | 3.8221E-01 | 2.0116E-01 | 1.0587E-01 | 6.9714E-02 | 4.8941E-02 | 4.0021E-02 | 3.5985E-02 |
| 0.20 | 1.1110E 00 | B | 1.0665E 00 | 9.2979E-01 | 7.5958E-01 | 6.0463E-01 | 4.9669E-01 | 4.3140E-01 | 3.9866E-01 |
| | | | 1.1813E 00 | 2.7228E 00 | 2.4230E 00 | 1.8242E 00 | 1.4363E 00 | 1.2172E 00 | 1.1114E 00 |
| | | H | 5.0121E-01 | 6.0143E-01 | 8.8982E-01 | 1.0313E 00 | 1.0168E 00 | 9.6627E-01 | 9.3001E-01 |
| | | | 4.6325E-01 | 3.3519E-01 | 2.0345E-01 | 1.3576E-01 | 1.0161E-01 | 8.4057E-02 | 7.5937E-02 |
| 0.30 | 9.4929E-01 | B | 9.1614E-01 | 8.0552E-01 | 6.5122E-01 | 5.1078E-01 | 4.1550E-01 | 3.5886E-01 | 3.3074E-01 |
| | | | 9.2877E-01 | 1.8031E 00 | 2.0224E 00 | 1.6790E 00 | 1.3817E 00 | 1.1962E 00 | 1.1029E 00 |
| | | H | 5.7748E-01 | 7.0259E-01 | 1.0388E 00 | 1.1449E 00 | 1.0929E 00 | 1.0207E 00 | 9.7443E-01 |
| | | | 5.3817E-01 | 4.3714E-01 | 2.9442E-01 | 2.0557E-01 | 1.5716E-01 | 1.3139E-01 | 1.1927E-01 |
| 0.40 | 8.2624E-01 | B | 7.9944E-01 | 7.0302E-01 | 5.6611E-01 | 4.2738E-01 | 3.4343E-01 | 2.9464E-01 | 2.7068E-01 |
| | | | 8.5288E-01 | 1.3045E 00 | 1.6964E 00 | 1.5348E 00 | 1.3157E 00 | 1.1624E 00 | 1.0813E 00 |
| | | H | 6.5909E-01 | 8.3192E-01 | 1.2165E 00 | 1.2665E 00 | 1.1689E 00 | 1.0724E 00 | 1.0153E 00 |
| | | | 6.1664E-01 | 5.2599E-01 | 3.8152E-01 | 2.7675E-01 | 2.1560E-01 | 1.8198E-01 | 1.6592E-01 |
| 0.50 | 7.2519E-01 | B | 7.0218E-01 | 6.1220E-01 | 4.6766E-01 | 3.5019E-01 | 2.7738E-01 | 2.3615E-01 | 2.1617E-01 |
| | | | 7.4991E-01 | 1.0154E 00 | 1.4321E 00 | 1.3966E 00 | 1.2437E 00 | 1.1202E 00 | 1.0516E 00 |
| | | H | 7.4991E-01 | 1.0154E 00 | 1.4321E 00 | 1.3966E 00 | 1.2437E 00 | 1.1202E 00 | 1.0516E 00 |
| | | | 7.0219E-01 | 6.1220E-01 | 4.6766E-01 | 3.5019E-01 | 2.7738E-01 | 2.3615E-01 | 2.1617E-01 |
| 0.60 | 6.3735E-01 | B | 6.1646E-01 | 5.2598E-01 | 3.8152E-01 | 2.7675E-01 | 2.1560E-01 | 1.8198E-01 | 1.6592E-01 |
| | | | 6.5909E-01 | 8.3192E-01 | 1.2165E 00 | 1.2665E 00 | 1.1689E 00 | 1.0724E 00 | 1.0153E 00 |
| | | H | 8.5388E-01 | 1.3045E 00 | 1.6964E 00 | 1.5348E 00 | 1.3157E 00 | 1.1962E 00 | 1.0813E 00 |
| | | | 7.9944E-01 | 7.0302E-01 | 5.6611E-01 | 4.2738E-01 | 3.4333E-01 | 2.9464E-01 | 2.7068E-01 |
| 0.70 | 5.5754E-01 | B | 5.3817E-01 | 4.3714E-01 | 2.9442E-01 | 2.0557E-01 | 1.5716E-01 | 1.3139E-01 | 1.1927E-01 |
| | | | 5.7748E-01 | 7.0259E-01 | 1.0388E 00 | 1.1449E 00 | 1.0929E 00 | 1.0207E 00 | 9.7443E-01 |
| | | H | 9.2877E-01 | 1.8031E 00 | 2.0224E 00 | 1.6790E 00 | 1.3817E 00 | 1.1962E 00 | 1.1114E 00 |
| | | | 9.1614E-01 | 8.0552E-01 | 6.5122E-01 | 5.1078E-01 | 4.1550E-01 | 3.5886E-01 | 3.3074E-01 |
| 0.80 | 4.8221E-01 | B | 4.6325E-01 | 3.3519E-01 | 2.0345E-01 | 1.3576E-01 | 1.0161E-01 | 8.4057E-02 | 7.5937E-02 |
| | | | 5.0121E-01 | 6.0143E-01 | 8.8982E-01 | 1.0313E 00 | 1.0168E 00 | 9.6627E-01 | 9.3001E-01 |
| | | H | 1.1813E 00 | 2.7228E 00 | 2.4230E 00 | 1.8242E 00 | 1.4363E 00 | 1.2172E 00 | 1.1114E 00 |
| | | | 1.0665E 00 | 9.2979E-01 | 7.5958E-01 | 6.0463E-01 | 4.9669E-01 | 4.3140E-01 | 3.9866E-01 |
| 0.90 | 4.0804E-01 | B | 3.8221E-01 | 2.0116E-01 | 1.0587E-01 | 6.9744E-02 | 4.8941E-02 | 4.0021E-02 | 3.5985E-02 |
| | | | 4.2701E-01 | 5.1472E-01 | 7.6215E-01 | 9.2487E-01 | 9.4086E-01 | 9.0966E-01 | 8.8267E-01 |
| | | H | 2.2481E 00 | 4.4915E 00 | 2.9055E 00 | 1.9580E 00 | 1.4692E 00 | 1.2166E 00 | 1.0994E 00 |
| | | | 1.2885E 00 | 1.0970E 00 | 8.9439E-01 | 7.1770E-01 | 5.9356E-01 | 5.1767E-01 | 4.7938E-01 |
| 1.00 | 3.2264E-01 | A | 0. | 0. | 0. | 0. | 0. | 0. | 0. |
| | | | 3.4694E-01 | 4.3283E-01 | 6.4885E-01 | 8.2369E-01 | 8.6478E-01 | 8.5082E-01 | 8.3242E-01 |
| | | H | 3.9299E 01 | 7.7379E 00 | 3.3661E 00 | 2.0000E 00 | 1.4226E 00 | 1.1434E 00 | 1.0261E 00 |
| | | | 2.1563E 00 | 1.5137E 00 | 1.1600E 00 | 9.1559E-01 | 7.5399E-01 | 6.5685E-01 | 6.0809E-01 |

ALBEDO =0.9000    PHI    THICKNESS = 2.000M

| DEPTH | PHI | | 1 | 2 | 3 | 4 | 5 | 6 | 7 |
|---|---|---|---|---|---|---|---|---|---|
| 0. | INFINITE | B | 2.2177E 00 | 1.5891E 00 | 1.2649E 00 | 1.0549E 00 | 9.1140E-01 | 8.1966E-01 | 7.7154E-01 |
| | | H | 3.9299E-01 | 7.7379E-01 | 3.3661E 00 | 2.0000E 00 | 1.4226E 00 | 1.1484E 00 | 1.0261E-01 |
| | | | 1.8206E-01 | 2.2155E-01 | 2.8811E-01 | 3.8829E-02 | 4.7411E-01 | 5.2099E-01 | 5.4002E-01 |
| | | | 0. | 0. | 0. | 0. | 0. | 0. | 0. |
| 0.10 | 1.4371E 00 | B | 1.3637E 00 | 1.1857E 00 | 1.0154E 00 | 8.7355E-01 | 7.6511E-01 | 6.9240E-01 | 6.5347E-01 |
| | | H | 2.3156E 00 | 4.5271E 00 | 2.9242E 00 | 1.9698E 00 | 1.4779E 00 | 1.2237E 00 | 1.1058E 00 |
| | | | 2.2174E-01 | 2.5843E-01 | 3.2678E-01 | 4.3176E-01 | 5.1721E-01 | 5.4099E-01 | 5.7760E-01 |
| | | | 1.9954E-01 | 1.0523E-01 | 5.5402E-02 | 3.5051E-02 | 2.5615E-02 | 2.0947E-02 | 1.8824E-02 |
| 0.20 | 1.1955E 00 | B | 1.1541E 00 | 1.0311E 00 | 8.9783E-01 | 7.7772E-01 | 6.8253E-01 | 6.1788E-01 | 5.8312E-01 |
| | | H | 1.2631E 00 | 2.7818E 00 | 2.4588E 00 | 1.8481E 00 | 1.4542E 00 | 1.2320E 00 | 1.1248E 00 |
| | | | 2.5711E-01 | 2.9367E-01 | 3.6570E-01 | 4.7664E-01 | 5.6110E-01 | 6.0125E-01 | 6.1513E-01 |
| | | | 2.3919E-01 | 1.7396E-01 | 1.0574E-01 | 7.0594E-02 | 5.2845E-02 | 4.3723E-02 | 3.9501E-02 |
| 0.30 | 1.0461E 00 | B | 1.0162E 00 | 9.2099E-01 | 8.0865E-01 | 7.0216E-01 | 6.1592E-01 | 5.5701E-01 | 5.2535E-01 |
| | | H | 1.0814E 00 | 1.8795E 00 | 2.0739E 00 | 1.7150E 00 | 1.4002E 00 | 1.2193E 00 | 1.1239E 00 |
| | | | 2.9182E-01 | 3.2899E-01 | 4.0613E-01 | 5.2324E-01 | 6.0633E-01 | 6.4213E-01 | 6.5293E-01 |
| | | | 2.7416E-01 | 2.2455E-01 | 1.5172E-01 | 1.0606E-01 | 8.1118E-02 | 6.7835E-02 | 6.1584E-02 |
| 0.40 | 9.3592E-01 | B | 9.1237E-01 | 8.3381E-01 | 7.3541E-01 | 6.3834E-01 | 5.5875E-01 | 5.0439E-01 | 4.7519E-01 |
| | | H | 9.6219E-01 | 1.3998E 00 | 1.7629E 00 | 1.5830E 00 | 1.3533E 00 | 1.1941E 00 | 1.1106E 00 |
| | | | 3.2674E-01 | 3.6504E-01 | 4.4385E-01 | 5.7271E-01 | 6.5317E-01 | 6.8376E-01 | 6.9106E-01 |
| | | | 3.0890E-01 | 2.6658E-01 | 1.9448E-01 | 1.4138E-01 | 1.1024E-01 | 9.3098E-02 | 8.4898E-02 |
| 0.50 | 8.4813E-01 | B | 8.2870E-01 | 7.6104E-01 | 6.7235E-01 | 5.8218E-01 | 5.0787E-01 | 4.5729E-01 | 4.3022E-01 |
| | | H | 8.6940E-01 | 1.1206E 00 | 1.5128E 00 | 1.4572E 00 | 1.2918E 00 | 1.1611E 00 | 1.0890E 00 |
| | | | 3.6240E-01 | 4.0228E-01 | 4.9464E-01 | 6.2538E-01 | 7.0181E-01 | 7.2620E-01 | 7.2953E-01 |
| | | | 3.4413E-01 | 3.0488E-01 | 2.3505E-01 | 1.7660E-01 | 1.4016E-01 | 1.1942E-01 | 1.0936E-01 |
| 0.60 | 7.7466E-01 | B | 7.5899E-01 | 6.9783E-01 | 6.1672E-01 | 5.3137E-01 | 4.6149E-01 | 4.1423E-01 | 3.8907E-01 |
| | | H | 7.9267E-01 | 9.5098E-01 | 1.3119E 00 | 1.3398E 00 | 1.2274E 00 | 1.1230E 00 | 1.0618E 00 |
| | | | 3.9923E-01 | 4.4114E-01 | 5.4405E-01 | 6.8176E-01 | 7.5240E-01 | 7.6942E-01 | 7.6827E-01 |
| | | | 3.8032E-01 | 3.4194E-01 | 2.7429E-01 | 2.1200E-01 | 1.7087E-01 | 1.4679E-01 | 1.3494E-01 |
| 0.70 | 7.1075E-01 | B | 6.9585E-01 | 6.4142E-01 | 5.6521E-01 | 4.8449E-01 | 4.1847E-01 | 3.7426E-01 | 3.5087E-01 |
| | | H | 7.2653E-01 | 9.5093E-01 | 1.1482E 00 | 1.2312E 00 | 1.1633E 00 | 1.0814E 00 | 1.0306E 00 |
| | | | 4.3759E-01 | 4.9209E-01 | 5.9921E-01 | 7.4240E-01 | 8.0505E-01 | 8.1339E-01 | 8.0718E-01 |
| | | | 4.1785E-01 | 3.7911E-01 | 3.1295E-01 | 2.4778E-01 | 2.0244E-01 | 1.7523E-01 | 1.6165E-01 |
| 0.80 | 6.5367E-01 | B | 6.4017E-01 | 5.9009E-01 | 5.1795E-01 | 4.4062E-01 | 3.7812E-01 | 3.3678E-01 | 3.1506E-01 |
| | | H | 6.7846E-01 | 7.5193E-01 | 1.0144E 00 | 1.1314E 00 | 1.0994E 00 | 1.0381E 00 | 9.9650E-01 |
| | | | 4.7785E-01 | 5.2568E-01 | 6.0455E-01 | 8.0789E-01 | 8.5981E-01 | 8.5799E-01 | 8.4610E-01 |
| | | | 4.5710E-01 | 4.1721E-01 | 3.5167E-01 | 2.8407E-01 | 2.3498E-01 | 2.0478E-01 | 1.8954E-01 |
| 0.90 | 6.0175E-01 | B | 5.8934E-01 | 5.4272E-01 | 4.7359E-01 | 3.9909E-01 | 3.3992E-01 | 3.0136E-01 | 2.8127E-01 |
| | | H | 6.1468E-01 | 6.2635E-01 | 9.0338E-01 | 1.0399E 00 | 1.0367E 00 | 9.9352E-01 | 9.6045E-01 |
| | | | 5.2044E-01 | 5.7270E-01 | 7.2989E-01 | 8.7884E-01 | 9.1672E-01 | 9.0305E-01 | 8.8483E-01 |
| | | | 4.9846E-01 | 5.5684E-01 | 3.9100E-01 | 3.2118E-01 | 2.6861E-01 | 2.3557E-01 | 2.1869E-01 |
| 1.00 | 5.5388E-01 | B | 5.4234E-01 | 4.9851E-01 | 4.3146E-01 | 3.5941E-01 | 3.0352E-01 | 2.6770E-01 | 2.4922E-01 |
| | | H | 5.6984E-01 | 6.2434E-01 | 8.0986E-01 | 9.5595E-01 | 9.7572E-01 | 9.4834E-01 | 9.2507E-01 |
| | | | 2.6384E-01 | 6.2434E-01 | 8.0586E-01 | 9.5599E-01 | 9.7572E-01 | 4.4834E-01 | 9.2307E-01 |
| | | | 5.4334E-01 | 4.9251E-01 | 4.3246E-01 | 3.5941E-01 | 3.0352E-01 | 2.6770E-01 | 2.4922E-01 |
| 1.10 | 5.0927E-01 | B | 4.9464E-01 | 4.6684E-01 | 3.9100E-01 | 3.2118E-01 | 2.6861E-01 | 2.3557E-01 | 2.1869E-01 |
| | | | 5.2044E-01 | 5.7270E-01 | 1.2497E-01 | 8.7884E-01 | 9.1672E-01 | 9.0305E-01 | 8.3483E-01 |

| μ | | c1 | c2 | c3 | c4 | c5 | c6 | c7 |
|---|---|---|---|---|---|---|---|---|
| 1.20 | 4.6733E-01 | 6.1468E-01 | 6.8261E-01 | 9.0338E-01 | 1.0399E 00 | 1.0367E 00 | 9.9352E-01 | 9.6045E-01 |
|      |            | 5.8934E-01 | 5.4272E-01 | 4.7352E-01 | 3.9909E-01 | 3.3092E-01 | 3.0136E-01 | 2.8127E-01 |
| 1.30 | 4.2760E-01 | 4.5710E-01 | 4.1721E-01 | 3.5167E-01 | 1.0381E 00 | 8.5999E-01 | 9.9650E-01 | 8.4610E-01 |
|      |            | 4.7785E-01 | 5.2568E-01 | 6.6045E-01 | 8.0739E-01 | 2.0478E-01 | 3.1506E-01 | 1.8934E-01 |
| 1.40 | 3.8968E-01 | 4.1785E-01 | 4.8209E-01 | 3.7911E-01 | 1.0816E 00 | 8.1339E-01 | 1.0306E 00 | 8.0718E-01 |
|      |            | 4.3759E-01 | 8.4622E-01 | 6.6412E-01 | 7.4240E-01 | 1.7523E-01 | 3.5087E-01 | 1.6165E-01 |
| 1.50 | 3.5320E-01 | 3.8032E-01 | 4.4114E-01 | 3.4194E-01 | 1.1230E 00 | 7.6942E-01 | 1.0618E 00 | 7.6827E-01 |
|      |            | 3.9923E-01 | 9.5098E-01 | 6.9783E-01 | 6.8176E-01 | 1.4679E-01 | 3.8907E-01 | 1.3494E-01 |
| 1.60 | 3.1778E-01 | 3.4413E-01 | 4.0228E-01 | 3.0488E-01 | 1.1611E 00 | 7.2620E-01 | 1.0890E 00 | 7.2993E-01 |
|      |            | 3.6240E-01 | 1.1206E 00 | 7.6103E-01 | 6.2538E-01 | 1.1942E-01 | 4.3022E-01 | 1.0936E-01 |
| 1.70 | 2.8299E-01 | 3.0890E-01 | 3.6504E-01 | 2.6658E-01 | 1.1941E 00 | 6.8376E-01 | 1.1106E 00 | 6.9106E-01 |
|      |            | 3.2674E-01 | 1.3358E 00 | 8.3391E-01 | 5.7271E-01 | 9.5098E-02 | 4.7519E-01 | 8.4898E-02 |
| 1.80 | 2.4824E-01 | 2.7416E-01 | 3.2899E-01 | 2.2455E-01 | 1.2193E 00 | 6.4213E-01 | 1.1258E 00 | 6.5293E-01 |
|      |            | 2.9132E-01 | 1.5795E 00 | 2.0999E 00 | 5.2324E-01 | 6.7835E-02 | 5.2556E-01 | 6.1584E-02 |
| 1.90 | 2.1253E-01 | 2.3919E-01 | 1.0844E 00 | 1.0523E-01 | 1.2320E 00 | 6.0125E-01 | 1.1248E 00 | 6.1513E-01 |
|      |            | 2.5711E-01 | 1.0162E 00 | 2.6849E-01 | 4.7644E-01 | 4.3723E-02 | 5.8312E-01 | 3.9301E-02 |
| 2.00 | 1.6077E-01 | 1.9956E-01 | 1.2631E 00 | 1.0206E-01 | 1.2237E 00 | 5.6099E-01 | 1.1058E 00 | 5.7760E-01 |
|      |            | 2.2174E-01 | 1.1541E 00 | 3.9299E-01 | 4.3176E-01 | 2.0947E-02 | 6.5347E-01 | 1.8824E-02 |

ALØSNG = 0.9000   THICKNESS = 3.0000

| DEPTH | PHI | | 1 | 2 | 3 | 4 | 5 | 6 | 7 |
|---|---|---|---|---|---|---|---|---|---|
| 0. | INFINITE | B | 2.2362E 00 | 1.6115E 00 | 1.2937E 00 | 1.0933E 00 | 9.5929E-01 | 8.7373E-01 | 8.2853E-01 |
| | | H | 3.9299E-01 | 7.7379E 00 | 1.3601E 00 | 2.0000E 00 | 1.4226E 00 | 1.1484E 00 | 1.0261E 00 |
| | | | 1.0387E-01 | 1.2595E-01 | 1.6653E-01 | 2.1036E-01 | 2.6743E-01 | 3.1147E-01 | 3.3511E-01 |
| | | | 0. | 0. | 0. | 0. | 0. | 0. | 0. |
| 0.50 | 8.8371E-01 | B | 8.6519E-01 | 3.0141E-01 | 7.2072E-01 | 6.4305E-01 | 5.7874E-01 | 5.3320E-01 | 5.0799E-01 |
| | | H | 9.0407E-01 | 1.1513E 00 | 1.5366E 00 | 1.4751E 00 | 1.3059E 00 | 1.1732E 00 | 1.1001E 00 |
| | | | 2.0396E-01 | 2.2513E-01 | 2.6437E-03 | 3.7775E-01 | 3.9504E-01 | 4.4026E-01 | 4.6192E-01 |
| | | | 1.9401E-01 | 1.7230E-01 | 1.3303E-01 | 1.0004E-01 | 7.9394E-02 | 6.7658E-02 | 6.1959E-02 |
| 1.00 | 6.0890E-01 | B | 5.9068E-01 | 5.5973E-01 | 5.0604E-01 | 4.5033E-01 | 4.0242E-01 | 3.6832E-01 | 3.4952E-01 |
| | | H | 6.1076E-01 | 6.7406E-01 | 8.5301E-01 | 9.9194E-01 | 1.0061E 00 | 9.7518E-01 | 9.4806E-01 |
| | | | 3.1002E-01 | 3.3576E-02 | 3.9831E-01 | 4.7593E-01 | 5.5169E-01 | 5.9164E-01 | 6.0691E-01 |
| | | | 2.9631E-01 | 2.7572E-01 | 2.3683E-01 | 2.0034E-01 | 1.6942E-01 | 1.4953E-01 | 1.3925E-01 |
| 1.50 | 4.3405E-01 | B | 4.2362E-01 | 3.9833E-01 | 3.5709E-01 | 3.1203E-01 | 2.7359E-01 | 2.4695E-01 | 2.3262E-01 |
| | | H | 4.4109E-01 | 4.7626E-01 | 5.6016E-01 | 6.5241E-01 | 7.5130E-01 | 7.7186E-01 | 7.7188E-01 |
| | | | 4.4169E-01 | 4.7592E-01 | 5.6016E-01 | 6.9241E-01 | 7.5170E-01 | 7.7116E-01 | 7.7188E-01 |
| | | | 4.2263E-01 | 3.9833E-01 | 3.5709E-01 | 3.1203E-01 | 2.7355E-01 | 2.4695E-01 | 2.3262E-01 |
| 2.00 | 3.0410E-01 | B | 2.9821E-01 | 2.7572E-01 | 2.3583E-01 | 2.0034E-01 | 1.6942E-01 | 1.4953E-01 | 1.3925E-01 |
| | | H | 3.1002E-01 | 3.3576E-01 | 3.8931E-01 | 4.7593E-01 | 5.5163E-01 | 5.5164E-01 | 6.0691E-01 |
| | | | 6.1976E-01 | 6.7406E-01 | 8.5301E-01 | 9.9194E-01 | 1.0061E 00 | 9.7518E-01 | 9.4806E-01 |
| | | | 5.9068E-01 | 5.5973E-01 | 5.0604E-01 | 4.5033E-01 | 4.0242E-01 | 3.5832E-01 | 3.4952E-01 |
| 2.50 | 1.9856E-01 | B | 1.9401E-01 | 1.7230E-01 | 1.3303E-01 | 1.0004E-01 | 7.9394E-02 | 6.7658E-02 | 6.1959E-02 |
| | | H | 2.0396E-01 | 2.2513E-01 | 2.6437E-01 | 3.2775E-01 | 3.9504E-01 | 4.4026E-01 | 4.6102E-01 |
| | | | 9.0407E-01 | 1.1513E 00 | 1.5366E 00 | 1.4751E 00 | 1.3059E 00 | 1.1732E 00 | 1.1001E 00 |
| | | | 8.6519E-01 | 8.0141E-01 | 7.2072E-01 | 6.4305E-01 | 5.7874E-01 | 5.3320E-01 | 5.0799E-01 |
| 3.00 | 9.5860E-02 | B | 0. | 0. | 0. | 0. | 0. | 0. | 0. |
| | | H | 3.9299E-01 | 7.7379E 00 | 1.3601E 00 | 2.0000E 00 | 1.4226E 00 | 1.1434E 00 | 1.0261E 00 |
| | | | 2.2362E 00 | 1.6115E 00 | 1.2937E 00 | 1.0933E 00 | 8.5929E-01 | 8.7373E-01 | 8.2853E-01 |

ALBEDO =0.9000   THICKNESS = 5.0000

| DEPTH | PHI | | 1 | 2 | 3 | 4 | 5 | 6 | 7 |
|---|---|---|---|---|---|---|---|---|---|
| 0. | INFINITE | B | 2.2444E 00 | 1.6216E 00 | 1.3065E 00 | 1.1099E 00 | 9.8098E-01 | 8.9877E-01 | 8.5590E-01 |
| | | | 3.9299E 01 | 7.7379E 00 | 3.3661E 00 | 2.0000E 00 | 1.4226E 00 | 1.1484E 00 | 1.0261E 00 |
| | | H | 3.5643E-02 | 4.3186E-02 | 5.4793E-02 | 7.0476E-02 | 8.9992E-02 | 1.0932E-01 | 1.2224E-01 |
| | | | 0. | 0. | 0. | 0. | | 0. | 0. |
| 0.50 | 8.9949E-01 | B | 8.8136E-01 | 8.1924E-01 | 7.4157E-01 | 6.6858E-01 | 6.0981E-01 | 5.6870E-01 | 5.4593E-01 |
| | | | 9.1946E-01 | 1.1650E 00 | 1.5471E 00 | 1.4830E 00 | 1.3122E 00 | 1.1785E 00 | 1.0150E 00 |
| | | H | 6.9660E-02 | 7.6752E-02 | 8.9395E-02 | 1.0796E-01 | 1.3202E-01 | 1.5548E-01 | 1.7066E-01 |
| | | | 6.6304E-02 | 5.8936E-02 | 4.5530E-02 | 3.4249E-02 | 2.7183E-02 | 2.3166E-02 | 2.1215E-02 |
| 1.00 | 6.3292E-01 | B | 6.2296E-01 | 5.8620E-01 | 5.3632E-01 | 4.8686E-01 | 4.4569E-01 | 4.1630E-01 | 3.9966E-01 |
| | | | 6.4333E-01 | 6.9587E-01 | 8.7199E-01 | 1.0078E 00 | 1.0196E 00 | 9.8703E-01 | 9.5909E-01 |
| | | H | 1.0499E-01 | 1.1335E-01 | 1.2871E-01 | 1.5215E-01 | 1.8279E-01 | 2.1136E-01 | 2.2902E-01 |
| | | | 1.0114E-01 | 9.3637E-02 | 8.1581E-02 | 6.8210E-02 | 5.7711E-02 | 5.0949E-02 | 4.7452E-02 |
| 1.50 | 4.6807E-01 | B | 4.6123E-01 | 4.3546E-01 | 3.9957E-01 | 3.6319E-01 | 3.3239E-01 | 3.1017E-01 | 2.9767E-01 |
| | | | 4.7515E-01 | 5.0710E-01 | 5.8826E-01 | 7.0698E-01 | 7.7326E-01 | 7.9065E-01 | 7.9022E-01 |
| | | H | 1.4727E-01 | 1.5762E-01 | 1.7701E-01 | 2.0767E-01 | 2.4706E-01 | 2.8124E-01 | 3.0112E-01 |
| | | | 1.4255E-01 | 1.3352E-01 | 1.2010E-01 | 1.0518E-01 | 9.2326E-02 | 8.3406E-02 | 7.8589E-02 |
| 2.00 | 3.5101E-01 | B | 3.4598E-01 | 3.2689E-01 | 2.9995E-01 | 2.7222E-01 | 2.4841E-01 | 2.3111E-01 | 2.2137E-01 |
| | | | 3.5618E-01 | 3.7909E-01 | 4.2764E-01 | 5.1103E-01 | 5.3310E-01 | 3.6951E-01 | 6.3423E-01 |
| | | H | 2.0005E-01 | 2.1324E-01 | 2.3841E-01 | 2.7982E-01 | 3.3039E-01 | 3.6551E-01 | 3.9055E-01 |
| | | | 1.9408E-01 | 1.8276E-01 | 1.6640E-01 | 1.4882E-01 | 1.3339E-01 | 1.2229E-01 | 1.1614E-01 |
| 2.50 | 2.6377E-01 | B | 2.5996E-01 | 2.4542E-01 | 2.2467E-01 | 2.0255E-01 | 1.8404E-01 | 1.7028E-01 | 1.6256E-01 |
| | | | 2.6768E-01 | 2.8485E-01 | 3.1865E-01 | 3.7653E-01 | 4.3913E-01 | 4.8098E-01 | 5.0076E-01 |
| | | H | 2.6768E-01 | 2.8485E-01 | 3.1865E-01 | 3.7653E-01 | 4.3913E-01 | 4.8098E-01 | 5.0076E-01 |
| | | | 2.5996E-01 | 2.4542E-01 | 2.2467E-01 | 2.0255E-01 | 1.8404E-01 | 1.7028E-01 | 1.6256E-01 |
| 3.00 | 1.9703E-01 | B | 1.9408E-01 | 1.8276E-01 | 1.6640E-01 | 1.4882E-01 | 1.3339E-01 | 1.2229E-01 | 1.1614E-01 |
| | | | 2.0005E-01 | 2.1324E-01 | 2.3841E-01 | 2.7982E-01 | 3.3017E-01 | 3.6951E-01 | 3.9055E-01 |
| | | H | 3.5618E-01 | 3.7909E-01 | 4.2764E-01 | 5.1103E-01 | 5.8310E-01 | 6.2042E-01 | 6.3423E-01 |
| | | | 3.4598E-01 | 3.2689E-01 | 2.9995E-01 | 2.7222E-01 | 2.4841E-01 | 2.3111E-01 | 2.2137E-01 |
| 3.50 | 1.4488E-01 | B | 1.4255E-01 | 1.3352E-01 | 1.2010E-01 | 1.0518E-01 | 9.2326E-02 | 8.3406E-02 | 7.8589E-02 |
| | | | 1.4727E-01 | 1.5762E-01 | 1.7701E-01 | 2.0767E-01 | 2.4706E-01 | 2.8124E-01 | 3.0112E-01 |
| | | H | 4.7515E-01 | 5.0710E-01 | 5.8826E-01 | 7.0698E-01 | 7.7326E-01 | 7.9065E-01 | 7.9022E-01 |
| | | | 4.6123E-01 | 4.3546E-01 | 3.9957E-01 | 3.6319E-01 | 3.3239E-01 | 3.1017E-01 | 2.9767E-01 |
| 4.00 | 1.0305E-01 | B | 1.0114E-01 | 9.3637E-02 | 8.1581E-02 | 6.8210E-02 | 5.7711E-02 | 5.0949E-02 | 4.7452E-02 |
| | | | 1.0499E-01 | 1.1335E-01 | 1.2871E-01 | 1.5215E-01 | 1.8279E-01 | 2.1136E-01 | 2.2902E-01 |
| | | H | 6.4333E-01 | 6.9587E-01 | 8.7199E-01 | 1.0078E 00 | 1.0196E 00 | 9.8703E-01 | 9.5909E-01 |
| | | | 6.2296E-01 | 5.8620E-01 | 5.3632E-01 | 4.8686E-01 | 4.4569E-01 | 4.1630E-01 | 3.9986E-01 |
| 4.50 | 6.7976E-02 | B | 6.6304E-02 | 5.8936E-02 | 4.5530E-02 | 3.4249E-02 | 2.7183E-02 | 2.3166E-02 | 2.1215E-02 |
| | | | 6.9605E-02 | 7.6752E-02 | 8.9395E-02 | 1.0796E-01 | 1.3202E-01 | 1.5548E-01 | 1.7066E-01 |
| | | H | 9.1946E-01 | 1.1650E 00 | 1.5471E 00 | 1.4830E 00 | 1.3122E 00 | 1.1785E 00 | 1.1050E 00 |
| | | | 8.8136E-01 | 8.1924E-01 | 7.4157E-01 | 6.6858E-01 | 6.0981E-01 | 5.6870E-01 | 5.4593E-01 |
| 5.00 | 3.3264E-02 | B | 0. | 0. | 0. | 7.0476E-02 | 0. | 0. | 0. |
| | | H | 3.5643E-02 | 4.3186E-02 | 5.4793E-02 | 2.0000E 00 | 8.9992E-02 | 1.0932E-01 | 1.2224E-01 |
| | | | 3.9299E 01 | 7.7379E 00 | 3.3661E 00 | 1.1099E 00 | 1.4226E 00 | 1.1484E 00 | 1.0261E 00 |
| | | | 2.2444E 00 | 1.6216E 00 | 1.3065E 00 | | 9.8035E-01 | 8.9877E-01 | 8.5590E-01 |

ALBEDO =1.0000    THICKNESS = 1.0000

| DEPTH | PHI | | 1 | 2 | 3 | 4 | 5 | 6 | 7 |
|---|---|---|---|---|---|---|---|---|---|
| 0. | INFINITE | B | 2.5297E 00 | 1.8316E 00 | 1.4365E 00 | 1.1481E 00 | 9.5143E-01 | 8.3152E-01 | 7.7094E-01 |
| | | B | 3.9299E 01 | 7.7379E 00 | 3.3661E 00 | 2.0000E 00 | 1.4226E 00 | 1.1484E 00 | 1.0261E 00 |
| | | H | 4.7409E-01 | 5.8909E-01 | 8.2889E-01 | 1.0001E 00 | 1.0262E 00 | 9.9908E-01 | 9.7300E-01 |
| | | H | 0. | 0. | 0. | 0. | 0. | 0. | 0. |
| 0.10 | 1.6703E 00 | B | 1.5884E 00 | 1.3801E 00 | 1.1439E 00 | 9.2583E-01 | 7.6892E-01 | 6.7203E-01 | 6.2292E-01 |
| | | B | 2.5581E 00 | 4.6714E 00 | 3.0020E 00 | 2.0194E 00 | 1.5142E 00 | 1.2534E 00 | 1.1326E 00 |
| | | H | 5.8559E-01 | 6.9738E-01 | 9.6025E-01 | 1.1103E 00 | 1.1063E 00 | 1.0595E 00 | 1.0238E 00 |
| | | H | 5.2398E-01 | 2.7551E-01 | 1.4497E-01 | 9.1701E-02 | 6.7009E-02 | 5.4794E-02 | 4.9242E-02 |
| 0.20 | 1.3985E 00 | B | 1.3502E 00 | 1.1970E 00 | 9.9050E-01 | 7.9336E-01 | 6.5365E-01 | 5.6856E-01 | 5.2576E-01 |
| | | B | 1.4733E 00 | 2.9644E 00 | 2.5762E 00 | 1.9200E 00 | 1.5145E 00 | 1.2820E 00 | 1.1701E 00 |
| | | H | 6.8520E-01 | 8.0657E-01 | 1.1014E 00 | 1.2214E 00 | 1.1823E 00 | 1.1143E 00 | 1.0686E 00 |
| | | H | 6.3466E-01 | 4.5932E-01 | 2.7877E-01 | 1.8601E-01 | 1.3921E-01 | 1.1517E-01 | 1.0404E-01 |
| 0.30 | 1.2222E 00 | B | 1.1852E 00 | 1.0564E 00 | 8.6224E-01 | 6.7920E-01 | 5.5359E-01 | 4.7860E-01 | 4.4128E-01 |
| | | B | 1.2644E 00 | 2.0650E 00 | 2.2118E 00 | 1.8147E 00 | 1.4866E 00 | 1.2845E 00 | 1.1832E 00 |
| | | H | 7.8405E-01 | 9.2744E-01 | 1.2594E 00 | 1.3352E 00 | 1.2548E 00 | 1.1636E 00 | 1.1075E 00 |
| | | H | 7.3355E-01 | 5.9742E-01 | 4.0263E-01 | 2.8118E-01 | 2.1497E-01 | 1.7974E-01 | 1.6316E-01 |
| 0.40 | 1.0838E 00 | B | 1.0528E 00 | 9.3587E-01 | 7.4525E-01 | 5.7428E-01 | 4.6205E-01 | 3.9664E-01 | 3.6449E-01 |
| | | B | 1.1174E 00 | 1.5677E 00 | 1.9074E 00 | 1.6940E 00 | 1.4419E 00 | 1.2698E 00 | 1.1799E 00 |
| | | H | 8.8594E-01 | 1.0737E 00 | 1.4409E 00 | 1.4521E 00 | 1.3231E 00 | 1.2070E 00 | 1.1399E 00 |
| | | H | 8.3343E-01 | 7.1469E-01 | 5.1949E-01 | 3.7710E-01 | 2.9386E-01 | 2.4807E-01 | 2.2619E-01 |
| 0.50 | 9.6588E-01 | B | 9.3822E-01 | 8.2460E-01 | 6.3250E-01 | 4.7436E-01 | 3.7598E-01 | 3.2021E-01 | 2.9315E-01 |
| | | B | 9.9500E-01 | 1.2705E 00 | 1.6539E 00 | 1.5721E 00 | 1.3862E 00 | 1.2430E 00 | 1.1646E 00 |
| | | H | 9.9500E-01 | 1.2705E 00 | 1.6539E 00 | 1.5721E 00 | 1.3862E 00 | 1.2430E 00 | 1.1646E 00 |
| | | H | 9.3822E-01 | 8.2460E-01 | 6.3250E-01 | 4.7436E-01 | 3.7598E-01 | 3.2021E-01 | 2.9315E-01 |
| 0.60 | 8.5930E-01 | B | 8.3343E-01 | 7.1469E-01 | 5.1949E-01 | 3.7710E-01 | 2.9386E-01 | 2.4807E-01 | 2.2619E-01 |
| | | B | 8.8594E-01 | 1.0737E 00 | 1.4409E 00 | 1.4521E 00 | 1.3231E 00 | 1.2070E 00 | 1.1399E 00 |
| | | H | 1.1174E 00 | 1.5677E 00 | 1.9074E 00 | 1.6940E 00 | 1.4419E 00 | 1.2698E 00 | 1.1799E 00 |
| | | H | 1.0528E 00 | 9.3587E-01 | 7.4525E-01 | 5.7428E-01 | 4.6205E-01 | 3.9664E-01 | 3.6449E-01 |
| 0.70 | 7.5867E-01 | B | 7.3355E-01 | 5.9742E-01 | 4.0263E-01 | 2.8118E-01 | 2.1497E-01 | 1.7974E-01 | 1.6316E-01 |
| | | B | 7.8405E-01 | 9.2744E-01 | 1.2594E 00 | 1.3352E 00 | 1.2548E 00 | 1.1636E 00 | 1.1075E 00 |
| | | H | 1.2644E 00 | 2.0650E 00 | 2.2118E 00 | 1.8147E 00 | 1.4866E 00 | 1.2845E 00 | 1.1832E 00 |
| | | H | 1.1852E 00 | 1.0564E 00 | 8.6224E-01 | 6.7920E-01 | 5.5359E-01 | 4.7860E-01 | 4.4128E-01 |
| 0.80 | 6.6013E-01 | B | 6.3466E-01 | 4.5932E-01 | 2.7877E-01 | 1.8601E-01 | 1.3921E-01 | 1.1517E-01 | 1.0404E-01 |
| | | B | 6.8520E-01 | 8.0657E-01 | 1.1014E 00 | 1.2214E 00 | 1.1823E 00 | 1.1143E 00 | 1.0686E 00 |
| | | H | 1.4733E 00 | 2.9644E 00 | 2.5762E 00 | 1.9280E 00 | 1.5145E 00 | 1.2820E 00 | 1.1701E 00 |
| | | H | 1.3502E 00 | 1.1970E 00 | 9.9050E-01 | 7.9336E-01 | 6.5365E-01 | 5.6856E-01 | 5.2576E-01 |
| 0.90 | 5.5970E-01 | B | 5.2398E-01 | 2.7551E-01 | 1.4497E-01 | 9.1701E-02 | 6.7009E-02 | 5.4794E-02 | 4.9242E-02 |
| | | B | 5.8559E-01 | 6.9738E-01 | 9.6025E-01 | 1.1103E 00 | 1.1063E 00 | 1.0595E 00 | 1.0238E 00 |
| | | H | 2.5581E 00 | 4.6714E 00 | 3.0020E 00 | 2.0194E 00 | 1.5142E 00 | 1.2534E 00 | 1.1326E 00 |
| | | H | 1.5884E 00 | 1.3801E 00 | 1.1439E 00 | 9.2583E-01 | 7.6892E-01 | 6.7203E-01 | 6.2292E-01 |
| 1.00 | 4.3948E-01 | B | 0. | 0. | 0. | 0. | 0. | 0. | 0. |
| | | B | 4.7409E-01 | 5.8909E-01 | 8.2889E-01 | 1.0001E 00 | 1.0262E 00 | 9.9908E-01 | 9.7300E-01 |
| | | H | 3.9299E 01 | 7.7379E 00 | 3.3661E 00 | 2.0000E 00 | 1.4226E 00 | 1.1484E 00 | 1.0261E 00 |
| | | H | 2.5297E 00 | 1.8316E 00 | 1.4365E 00 | 1.1481E 00 | 9.5143E-01 | 8.3152E-01 | 7.7094E-01 |

ALBEDO =1.0000     THICKNESS = 2.0000

| DEPTH | PHI | | 1 | 2 | 3 | 4 | 5 | 6 | 7 |
|---|---|---|---|---|---|---|---|---|---|
| 0. | INFINITE | B | 2.6689E 00 | 2.0025E 00 | 1.6626E 00 | 1.4307E 00 | 1.2605E 00 | 1.1463E 00 | 1.0848E 00 |
| | | | 3.9299E 01 | 7.7379E 01 | 3.3661E 00 | 2.0000E 00 | 1.4226E 00 | 1.1484E 00 | 1.0261E 00 |
| | | H | 3.1899E-01 | 3.8889E-01 | 4.9319E-01 | 6.2111E-01 | 7.1620E-01 | 7.6196E-01 | 7.7800E-01 |
| | | | 0. | 0. | 0. | 0. | 0. | 0. | 0. |
| 0.10 | 1.8341E 00 | B | 1.7596E 00 | 1.5803E 00 | 1.4017E 00 | 1.2384E 00 | 1.1027E 00 | 1.0072E 00 | 9.5476E-01 |
| | | | 2.7117E 00 | 4.7522E 00 | 3.0445E 00 | 2.0463E 00 | 1.5339E 00 | 1.2695E 00 | 1.1470E 00 |
| | | H | 3.9078E-01 | 4.5359E-01 | 5.5599E-01 | 6.8399E-01 | 7.7410E-01 | 8.1360E-01 | 8.2557E-01 |
| | | | 3.5114E-01 | 1.8492E-01 | 9.7328E-02 | 6.1571E-02 | 4.4994E-02 | 3.6793E-02 | 3.3065E-02 |
| 0.20 | 1.5906E 00 | B | 1.5493E 00 | 1.4251E 00 | 1.2805E 00 | 1.1358E 00 | 1.0113E 00 | 9.2285E-01 | 8.7425E-01 |
| | | | 1.6583E 00 | 3.0986E 00 | 2.6577E 00 | 1.9824E 00 | 1.5552E 00 | 1.3157E 00 | 1.2005E 00 |
| | | H | 4.5284E-01 | 5.1347E-01 | 6.1658E-01 | 7.4569E-01 | 8.3043E-01 | 8.6323E-01 | 8.7093E-01 |
| | | | 4.2161E-01 | 3.0638E-01 | 1.8616E-01 | 1.2426E-01 | 9.3012E-02 | 7.6953E-02 | 6.9521E-02 |
| 0.30 | 1.4415E 00 | B | 1.4113E 00 | 1.3127E 00 | 1.1863E 00 | 1.0521E 00 | 9.3451E-01 | 8.5100E-01 | 8.0521E-01 |
| | | | 1.4768E 00 | 2.2387E 00 | 2.3290E 00 | 1.8967E 00 | 1.5493E 00 | 1.3368E 00 | 1.2308E 00 |
| | | H | 5.1200E-01 | 5.7158E-01 | 6.7690E-01 | 8.0750E-01 | 8.8602E-01 | 9.1146E-01 | 9.1458E-01 |
| | | | 4.8213E-01 | 3.9525E-01 | 2.6700E-01 | 1.8667E-01 | 1.4277E-01 | 1.1939E-01 | 1.0839E-01 |
| 0.40 | 1.3299E 00 | B | 1.3058E 00 | 1.2218E 00 | 1.1062E 00 | 9.7827E-01 | 8.6557E-01 | 7.8607E-01 | 7.4262E-01 |
| | | | 1.3567E 00 | 1.7741E 00 | 2.0579E 00 | 1.8033E 00 | 1.5271E 00 | 1.3418E 00 | 1.2456E 00 |
| | | H | 5.6965E-01 | 6.2884E-01 | 7.3784E-01 | 8.6998E-01 | 9.4113E-01 | 9.5840E-01 | 9.5659E-01 |
| | | | 5.4044E-01 | 4.6781E-01 | 3.4160E-01 | 2.4840E-01 | 1.9371E-01 | 1.6359E-01 | 1.4918E-01 |
| 0.50 | 1.2391E 00 | B | 1.2185E 00 | 1.1433E 00 | 1.0343E 00 | 9.1042E-01 | 8.0168E-01 | 7.2547E-01 | 6.8412E-01 |
| | | | 1.2613E 00 | 1.5065E 00 | 1.8357E 00 | 1.7087E 00 | 1.4946E 00 | 1.3353E 00 | 1.2492E 00 |
| | | H | 6.2657E-01 | 6.8586E-01 | 8.0018E-01 | 9.3358E-01 | 9.9588E-01 | 1.0041E 00 | 9.9694E-01 |
| | | | 5.9785E-01 | 5.3219E-01 | 4.1124E-01 | 3.0934E-01 | 2.4551E-01 | 2.0922E-01 | 1.9160E-01 |
| 0.60 | 1.1605E 00 | B | 1.1420E 00 | 1.0724E 00 | 9.6750E-01 | 8.4624E-01 | 7.4079E-01 | 6.6775E-01 | 6.2841E-01 |
| | | | 1.1799E 00 | 1.3382E 00 | 1.6531E 00 | 1.6161E 00 | 1.4553E 00 | 1.3205E 00 | 1.2442E 00 |
| | | H | 6.8324E-01 | 7.4308E-01 | 8.6472E-01 | 9.9869E-01 | 1.0503E 00 | 1.0484E 00 | 1.0355E 00 |
| | | | 6.5440E-01 | 5.9248E-01 | 4.7722E-01 | 3.6954E-01 | 2.9798E-01 | 2.5607E-01 | 2.3543E-01 |
| 0.70 | 1.0893E 00 | B | 1.0722E 00 | 1.0064E 00 | 9.0391E-01 | 7.8440E-01 | 6.8202E-01 | 6.1209E-01 | 5.7475E-01 |
| | | | 1.1071E 00 | 1.2206E 00 | 1.5014E 00 | 1.5268E 00 | 1.4113E 00 | 1.2991E 00 | 1.2325E 00 |
| | | H | 7.4006E-01 | 8.0094E-01 | 9.3240E-01 | 1.0657E 00 | 1.1044E 00 | 1.0912E 00 | 1.0721E 00 |
| | | | 7.1111E-01 | 6.5083E-01 | 5.4054E-01 | 4.2910E-01 | 3.5100E-01 | 3.0399E-01 | 2.8050E-01 |
| 0.80 | 1.0231E 00 | B | 1.0069E 00 | 1.0069E 00 | 8.4234E-01 | 7.2405E-01 | 6.2474E-01 | 5.5801E-01 | 5.2270E-01 |
| | | | 1.0397E 00 | 1.1296E 00 | 1.3736E 00 | 1.4412E 00 | 1.3641E 00 | 1.2727E 00 | 1.2152E 00 |
| | | H | 7.9737E-01 | 8.5994E-01 | 1.0040E 00 | 1.1348E 00 | 1.1581E 00 | 1.1323E 00 | 1.1066E 00 |
| | | | 7.6812E-01 | 7.0842E-01 | 6.0203E-01 | 4.8818E-01 | 4.0454E-01 | 3.5288E-01 | 3.2674E-01 |
| 0.90 | 9.6019E-01 | B | 9.4464E-01 | 8.8305E-01 | 7.8198E-01 | 6.6462E-01 | 5.6857E-01 | 5.0519E-01 | 4.7199E-01 |
| | | | 9.7603E-01 | 1.0530E 00 | 1.2638E 00 | 1.3595E 00 | 1.3146E 00 | 1.2421E 00 | 1.1933E 00 |
| | | H | 8.5550E-01 | 9.2076E-01 | 1.0822E 00 | 1.2066E 00 | 1.2112E 00 | 1.1714E 00 | 1.1385E 00 |
| | | | 8.2578E-01 | 7.6597E-01 | 6.6238E-01 | 5.4697E-01 | 4.5861E-01 | 4.0270E-01 | 3.7406E-01 |
| 1.00 | 8.9957E-01 | B | 8.8447E-01 | 8.2402E-01 | 7.2218E-01 | 6.0570E-01 | 5.1325E-01 | 4.5346E-01 | 4.2247E-01 |
| | | | 9.1488E-01 | 9.8449E-01 | 1.1679E 00 | 1.2814E 00 | 1.2634E 00 | 1.2081E 00 | 1.1676E 00 |
| | | H | 9.1488E-01 | 9.8449E-01 | 1.1679E 00 | 1.2814E 00 | 1.2634E 00 | 1.2081E 00 | 1.1676E 00 |
| | | | 8.8447E-01 | 8.2402E-01 | 7.2218E-01 | 6.0570E-01 | 5.1325E-01 | 4.5346E-01 | 4.2247E-01 |
| 1.10 | 8.4057E-01 | B | 8.2578E-01 | 7.6597E-01 | 6.6238E-01 | 5.4697E-01 | 4.5861E-01 | 4.0270E-01 | 3.7406E-01 |
| | | | 8.5550E-01 | 9.2076E-01 | 1.0822E 00 | 1.2066E 00 | 1.2112E 00 | 1.1714E 00 | 1.1385E 00 |

| | | | | | | | | | |
|---|---|---|---|---|---|---|---|---|---|
| | | H | 9.7603E-01 | 1.0530E 00 | 1.2638E 00 | 1.3595E 00 | 1.3146E 00 | 1.2421E 00 | 1.1933E 00 |
| | | | 9.4464E-01 | 8.8305E-01 | 7.8198E-01 | 6.6462E-01 | 5.6857E-01 | 5.0519E-01 | 4.7199E-01 |
| 1.20 | 7.8269E-01 | B | 7.6812E-01 | 7.0842E-01 | 6.0203E-01 | 4.8818E-01 | 4.0454E-01 | 3.5288E-01 | 3.2674E-01 |
| | | | 7.9737E-01 | 8.5994E-01 | 1.0044E 00 | 1.1348E 00 | 1.1581E 00 | 1.1323E 00 | 1.1066E 00 |
| | | H | 1.0397E 00 | 1.1296E 00 | 1.3736E 00 | 1.4412E 00 | 1.3641E 00 | 1.2727E 00 | 1.2152E 00 |
| | | | 1.0069E 00 | 9.4360E-01 | 8.4234E-01 | 7.2405E-01 | 6.2474E-01 | 5.5801E-01 | 5.2270E-01 |
| 1.30 | 7.2556E-01 | B | 7.1111E-01 | 6.5083E-01 | 5.4054E-01 | 4.2910E-01 | 3.5100E-01 | 3.0399E-01 | 2.8050E-01 |
| | | | 7.4006E-01 | 8.0094E-01 | 9.3240E-01 | 1.0657E 00 | 1.1044E 00 | 1.0912E 00 | 1.0721E 00 |
| | | H | 1.1071E 00 | 1.2206E 00 | 1.5014E 00 | 1.5268E 00 | 1.4113E 00 | 1.2991E 00 | 1.2325E 00 |
| | | | 1.0722E 00 | 1.0064E 00 | 9.0391E-01 | 7.8440E-01 | 6.8202E-01 | 6.1209E-01 | 5.7475E-01 |
| 1.40 | 6.6882E-01 | B | 6.5440E-01 | 5.9248E-01 | 4.7722E-01 | 3.6954E-01 | 2.9798E-01 | 2.5607E-01 | 2.3543E-01 |
| | | | 6.8324E-01 | 7.4308E-01 | 8.6472E-01 | 9.9869E-01 | 1.0503E 00 | 1.0484E 00 | 1.0355E 00 |
| | | H | 1.1799E 00 | 1.3382E 00 | 1.6531E 00 | 1.6161E 00 | 1.4553E 00 | 1.3205E 00 | 1.2442E 00 |
| | | | 1.1420E 00 | 1.0724E 00 | 9.6750E-01 | 8.4624E-01 | 7.4079E-01 | 6.6775E-01 | 6.2841E-01 |
| 1.50 | 6.1213E-01 | B | 5.9765E-01 | 5.3219E-01 | 4.1124E-01 | 3.0934E-01 | 2.4551E-01 | 2.0922E-01 | 1.9160E-01 |
| | | | 6.2657E-01 | 6.8586E-01 | 8.0018E-01 | 9.3358E-01 | 9.9588E-01 | 1.0041E 00 | 9.9694E-01 |
| | | H | 1.2613E 00 | 1.5065E 00 | 1.8357E 00 | 1.7087E 00 | 1.4946E 00 | 1.3353E 00 | 1.2492E 00 |
| | | | 1.2185E 00 | 1.1433E 00 | 1.0343E 00 | 9.1042E-01 | 8.0168E-01 | 7.2547E-01 | 6.8412E-01 |
| 1.60 | 5.5510E-01 | B | 5.4044E-01 | 4.6781E-01 | 3.4160E-01 | 2.4840E-01 | 1.9371E-01 | 1.6359E-01 | 1.4918E-01 |
| | | | 5.6965E-01 | 6.2884E-01 | 7.3784E-01 | 8.6998E-01 | 9.4113E-01 | 9.5840E-01 | 9.5659E-01 |
| | | H | 1.3567E 00 | 1.7741E 00 | 2.0579E 00 | 1.8033E 00 | 1.5271E 00 | 1.3418E 00 | 1.2456E 00 |
| | | | 1.3058E 00 | 1.2218E 00 | 1.1062E 00 | 9.7827E-01 | 8.6567E-01 | 7.8607E-01 | 7.4262E-01 |
| 1.70 | 4.9718E-01 | B | 4.8213E-01 | 3.9525E-01 | 2.6706E-01 | 1.8667E-01 | 1.4277E-01 | 1.1939E-01 | 1.0839E-01 |
| | | | 5.1200E-01 | 5.7158E-01 | 6.7690E-01 | 8.0750E-01 | 8.8602E-01 | 9.1146E-01 | 9.1458E-01 |
| | | H | 1.4768E 00 | 2.2387E 00 | 2.3290E 00 | 1.8967E 00 | 1.5493E 00 | 1.3368E 00 | 1.2308E 00 |
| | | | 1.4113E 00 | 1.3127E 00 | 1.1863E 00 | 1.0521E 00 | 9.3451E-01 | 8.5100E-01 | 8.0521E-01 |
| 1.80 | 4.3750E-01 | B | 4.2161E-01 | 3.0638E-01 | 1.8616E-01 | 1.2426E-01 | 9.3012E-02 | 7.6953E-02 | 6.9521E-02 |
| | | | 4.5284E-01 | 5.1347E-01 | 6.1658E-01 | 7.4569E-01 | 8.3043E-01 | 8.6323E-01 | 8.7093E-01 |
| | | H | 1.6583E 00 | 3.0986E 00 | 2.6577E 00 | 1.9824E 00 | 1.5552E 00 | 1.3157E 00 | 1.2005E 00 |
| | | | 1.5493E 00 | 1.4251E 00 | 1.2805E 00 | 1.1358E 00 | 1.0113E 00 | 9.2285E-01 | 8.7425E-01 |
| 1.90 | 3.7438E-01 | B | 3.5114E-01 | 1.8492E-01 | 9.7328E-02 | 6.1571E-02 | 4.4994E-02 | 3.6793E-02 | 3.3065E-02 |
| | | | 3.9078E-01 | 4.5359E-01 | 5.5599E-01 | 6.8399E-01 | 7.7410E-01 | 8.1360E-01 | 8.2557E-01 |
| | | H | 2.7117E 00 | 4.7522E 00 | 3.0445E 00 | 2.0463E 00 | 1.5339E 00 | 1.2695E 00 | 1.1470E 00 |
| | | | 1.7596E 00 | 1.5803E 00 | 1.4017E 00 | 1.2384E 00 | 1.1027E 00 | 1.0072E 00 | 9.5476E-01 |
| 2.00 | 2.9633E-01 | B | 0. | 0. | 0. | 0. | 0. | 0. | 0. |
| | | | 3.1899E-01 | 3.8869E-01 | 4.9319E-01 | 6.2111E-01 | 7.1620E-01 | 7.6196E-01 | 7.7800E-01 |
| | | H | 3.9299E 01 | 7.7379E 00 | 3.3661E 00 | 2.0000E 00 | 1.4226E 00 | 1.1484E 00 | 1.0261E 00 |
| | | | 2.6689E 00 | 2.0025E 00 | 1.6626E 00 | 1.4307E 00 | 1.2605E 00 | 1.1463E 00 | 1.0848E 00 |

ALBEDO =1.0000    THICKNESS = 5.0000

| DEPTH | PHI | | 1 | 2 | 3 | 4 | 5 | 6 | 7 |
|---|---|---|---|---|---|---|---|---|---|
| 0. | INFINITE | B | 2.8173E 00 | 2.1830E 00 | 1.8893E 00 | 1.7120E 00 | 1.5915E 00 | 1.5106E 00 | 1.4658E 00 |
|  |  |  | 3.9299E 01 | 7.7379E 00 | 3.3661E 00 | 2.0000E 00 | 1.4226E 00 | 1.1484E 00 | 1.0261E 00 |
|  |  | H | 1.6763E-01 | 2.0386E-01 | 2.5509E-01 | 3.1380E-01 | 3.7121E-01 | 4.1710E-01 | 4.4402E-01 |
|  |  |  | 0. | 0. | 0. | 0. | 0. | 0. | 0. |
| 1.00 | 1.3090E 00 | B | 1.3003E 00 | 1.2666E 00 | 1.2153E 00 | 1.1558E 00 | 1.0978E 00 | 1.0509E 00 | 1.0225E 00 |
|  |  |  | 1.3178E 00 | 1.3612E 00 | 1.4989E 00 | 1.5595E 00 | 1.4993E 00 | 1.4166E 00 | 1.3618E 00 |
|  |  | H | 4.6558E-01 | 4.9409E-01 | 5.4005E-01 | 5.9601E-01 | 6.5180E-01 | 6.9434E-01 | 7.1761E-01 |
|  |  |  | 4.5154E-01 | 4.2261E-01 | 3.7189E-01 | 3.1262E-01 | 2.6520E-01 | 2.3444E-01 | 2.1848E-01 |
| 2.00 | 1.0077E 00 | B | 1.0006E 00 | 9.7165E-01 | 9.2527E-01 | 8.6935E-01 | 8.1381E-01 | 7.6930E-01 | 7.4287E-01 |
|  |  |  | 1.0148E 00 | 1.0444E 00 | 1.0971E 00 | 1.1682E 00 | 1.2109E 00 | 1.2179E 00 | 1.2121E 00 |
|  |  | H | 7.3902E-01 | 7.6740E-01 | 8.1376E-01 | 8.7249E-01 | 9.2767E-01 | 9.6223E-01 | 9.7729E-01 |
|  |  |  | 7.2512E-01 | 6.9680E-01 | 6.5077E-01 | 5.9422E-01 | 5.3969E-01 | 4.9849E-01 | 4.7513E-01 |
| 3.00 | 7.3207E-01 | B | 7.2512E-01 | 6.9680E-01 | 6.5077E-01 | 5.9422E-01 | 5.3969E-01 | 4.9849E-01 | 4.7513E-01 |
|  |  |  | 7.3902E-01 | 7.6740E-01 | 8.1376E-01 | 8.7249E-01 | 9.2767E-01 | 9.6223E-01 | 9.7729E-01 |
|  |  | H | 1.0148E 00 | 1.0444E 00 | 1.0971E 00 | 1.1682E 00 | 1.2109E 00 | 1.2179E 00 | 1.2121E 00 |
|  |  |  | 1.0006E 00 | 9.7165E-01 | 9.2527E-01 | 8.6935E-01 | 8.1381E-01 | 7.6930E-01 | 7.4287E-01 |
| 4.00 | 4.5857E-01 | B | 4.5154E-01 | 4.2261E-01 | 3.7189E-01 | 3.1262E-01 | 2.6520E-01 | 2.3444E-01 | 2.1848E-01 |
|  |  |  | 4.6558E-01 | 4.9409E-01 | 5.4005E-01 | 5.9601E-01 | 6.5180E-01 | 6.9434E-01 | 7.1761E-01 |
|  |  | H | 1.3178E 00 | 1.3612E 00 | 1.4989E 00 | 1.5595E 00 | 1.4993E 00 | 1.4166E 00 | 1.3618E 00 |
|  |  |  | 1.3003E 00 | 1.2666E 00 | 1.2153E 00 | 1.1558E 00 | 1.0978E 00 | 1.0509E 00 | 1.0225E 00 |
| 5.00 | 1.5579E-01 | B | 0. | 0. | 0. | 0. | 0. | 0. | 0. |
|  |  |  | 1.6763E-01 | 2.0386E-01 | 2.5509E-01 | 3.1380E-01 | 3.7121E-01 | 4.1710E-01 | 4.4402E-01 |
|  |  | H | 3.9299E 01 | 7.7379E 00 | 3.3661E 00 | 2.0000E 00 | 1.4226E 00 | 1.1484E 00 | 1.0261E 00 |
|  |  |  | 2.8173E 00 | 2.1830E 00 | 1.8893E 00 | 1.7120E 00 | 1.5915E 00 | 1.5106E 00 | 1.4658E 00 |

ALBEDO =1.0000     THICKNESS = 10.0000

| DEPTH | PHI | | 1 | 2 | 3 | 4 | 5 | 6 | 7 |
|---|---|---|---|---|---|---|---|---|---|
| 0. | INFINITE | B | 2.8907E 00 | 2.2722E 00 | 2.0009E 00 | 1.8493E 00 | 1.7538E 00 | 1.6933E 00 | 1.6607E 00 |
| | | | 3.9299E 01 | 7.7379E 00 | 3.3661E 00 | 2.0000E 00 | 1.4226E 00 | 1.1484E 00 | 1.0261E 00 |
| | | H | 9.4209E-02 | 1.1457E-01 | 1.4333E-01 | 1.7624E-01 | 2.0832E-01 | 2.3454E-01 | 2.5065E-01 |
| | | | 0. | 0. | 0. | 0. | 0. | 0. | 0. |
| 1.00 | 1.5096E 00 | B | 1.5040E 00 | 1.4828E 00 | 1.4515E 00 | 1.4162E 00 | 1.3824E 00 | 1.3551E 00 | 1.3384E 00 |
| | | | 1.5154E 00 | 1.5461E 00 | 1.6617E 00 | 1.6963E 00 | 1.6154E 00 | 1.5192E 00 | 1.4574E 00 |
| | | H | 2.6153E-01 | 2.7751E-01 | 3.0324E-01 | 3.3423E-01 | 3.6516E-01 | 3.9070E-01 | 4.0646E-01 |
| | | | 2.5365E-01 | 2.3741E-01 | 2.0893E-01 | 1.7564E-01 | 1.4900E-01 | 1.3172E-01 | 1.2275E-01 |
| 2.00 | 1.3277E 00 | B | 1.3236E 00 | 1.3070E 00 | 1.2806E 00 | 1.2490E 00 | 1.2177E 00 | 1.1919E 00 | 1.1759E 00 |
| | | | 1.3318E 00 | 1.3490E 00 | 1.3817E 00 | 1.4281E 00 | 1.4470E 00 | 1.4359E 00 | 1.4199E 00 |
| | | H | 4.1455E-01 | 4.3033E-01 | 4.5584E-01 | 4.8666E-01 | 5.1748E-01 | 5.4295E-01 | 5.5862E-01 |
| | | | 4.0680E-01 | 3.9098E-01 | 3.6523E-01 | 3.3355E-01 | 3.0297E-01 | 2.7986E-01 | 2.6675E-01 |
| 3.00 | 1.1709E 00 | B | 1.1670E 00 | 1.1511E 00 | 1.1254E 00 | 1.0944E 00 | 1.0635E 00 | 1.0380E 00 | 1.0222E 00 |
| | | | 1.1748E 00 | 1.1908E 00 | 1.2172E 00 | 1.2518E 00 | 1.2828E 00 | 1.2980E 00 | 1.3016E 00 |
| | | H | 5.6649E-01 | 5.8224E-01 | 6.0771E-01 | 6.3851E-01 | 6.6933E-01 | 6.9474E-01 | 7.1027E-01 |
| | | | 5.5876E-01 | 5.4300E-01 | 5.1750E-01 | 4.8652E-01 | 4.5562E-01 | 4.3081E-01 | 4.1605E-01 |
| 4.00 | 1.0181E 00 | B | 1.0142E 00 | 9.9843E-01 | 9.7292E-01 | 9.4209E-01 | 9.1128E-01 | 8.8581E-01 | 8.7011E-01 |
| | | | 1.0220E 00 | 1.0378E 00 | 1.0634E 00 | 1.0950E 00 | 1.1265E 00 | 1.1492E 00 | 1.1604E 00 |
| | | H | 7.1823E-01 | 7.3398E-01 | 7.5945E-01 | 7.9027E-01 | 8.2116E-01 | 8.4644E-01 | 8.6158E-01 |
| | | | 7.1051E-01 | 6.9476E-01 | 6.6929E-01 | 6.3847E-01 | 6.0763E-01 | 5.8234E-01 | 5.6696E-01 |
| 5.00 | 8.6613E-01 | B | 8.6227E-01 | 8.4651E-01 | 8.2104E-01 | 7.9024E-01 | 7.5943E-01 | 7.3402E-01 | 7.1840E-01 |
| | | | 8.6999E-01 | 8.8575E-01 | 9.1126E-01 | 9.4221E-01 | 9.7331E-01 | 9.8044E-01 | 1.0121E 00 |
| | | H | 8.6999E-01 | 8.8575E-01 | 9.1126E-01 | 9.4221E-01 | 9.7331E-01 | 9.8044E-01 | 1.0121E 00 |
| | | | 8.6227E-01 | 8.4651E-01 | 8.2104E-01 | 7.9024E-01 | 7.5943E-01 | 7.3402E-01 | 7.1840E-01 |
| 6.00 | 7.1437E-01 | B | 7.1051E-01 | 6.9476E-01 | 6.6929E-01 | 6.3847E-01 | 6.0763E-01 | 5.8234E-01 | 5.6696E-01 |
| | | | 7.1823E-01 | 7.3398E-01 | 7.5945E-01 | 7.9027E-01 | 8.2116E-01 | 8.4644E-01 | 8.6158E-01 |
| | | H | 1.0220E 00 | 1.0378E 00 | 1.0634E 00 | 1.0950E 00 | 1.1265E 00 | 1.1492E 00 | 1.1604E 00 |
| | | | 1.0142E 00 | 9.9843E-01 | 9.7292E-01 | 9.4209E-01 | 9.1128E-01 | 8.8581E-01 | 8.7011E-01 |
| 7.00 | 5.6262E-01 | B | 5.5876E-01 | 5.4300E-01 | 5.1750E-01 | 4.8652E-01 | 4.5562E-01 | 4.3081E-01 | 4.1605E-01 |
| | | | 5.6649E-01 | 5.8224E-01 | 6.0771E-01 | 6.3851E-01 | 6.6933E-01 | 6.9474E-01 | 7.1027E-01 |
| | | H | 1.1748E 00 | 1.1908E 00 | 1.2172E 00 | 1.2518E 00 | 1.2828E 00 | 1.2980E 00 | 1.3016E 00 |
| | | | 1.1670E 00 | 1.1511E 00 | 1.1254E 00 | 1.0944E 00 | 1.0635E 00 | 1.0380E 00 | 1.0222E 00 |
| 8.00 | 4.1067E-01 | B | 4.0680E-01 | 3.9098E-01 | 3.6523E-01 | 3.3355E-01 | 3.0297E-01 | 2.7986E-01 | 2.6675E-01 |
| | | | 4.1455E-01 | 4.3033E-01 | 4.5584E-01 | 4.8666E-01 | 5.1748E-01 | 5.4295E-01 | 5.5862E-01 |
| | | H | 1.3318E 00 | 1.3490E 00 | 1.3817E 00 | 1.4281E 00 | 1.4470E 00 | 1.4359E 00 | 1.4199E 00 |
| | | | 1.3236E 00 | 1.3070E 00 | 1.2806E 00 | 1.2490E 00 | 1.2177E 00 | 1.1919E 00 | 1.1759E 00 |
| 9.00 | 2.5759E-01 | B | 2.5365E-01 | 2.3741E-01 | 2.0893E-01 | 1.7564E-01 | 1.4900E-01 | 1.3172E-01 | 1.2275E-01 |
| | | | 2.6153E-01 | 2.7751E-01 | 3.0324E-01 | 3.3423E-01 | 3.6516E-01 | 3.9070E-01 | 4.0646E-01 |
| | | H | 1.5154E 00 | 1.5461E 00 | 1.6617E 00 | 1.6963E 00 | 1.6154E 00 | 1.5192E 00 | 1.4574E 00 |
| | | | 1.5040E 00 | 1.4828E 00 | 1.4515E 00 | 1.4162E 00 | 1.3824E 00 | 1.3551E 00 | 1.3384E 00 |
| 10.00 | 8.7559E-02 | B | 9.4209E-02 | 1.1457E-01 | 1.4333E-01 | 1.7624E-01 | 2.0832E-01 | 2.3454E-01 | 2.5065E-01 |
| | | | 3.9299E 01 | 7.7379E 00 | 3.3661E 00 | 2.0000E 00 | 1.4226E 00 | 1.1484E 00 | 1.0261E 00 |
| | | H | 2.8907E 00 | 2.2722E 00 | 2.0009E 00 | 1.8493E 00 | 1.7538E 00 | 1.6933E 00 | 1.6607E 00 |

ALBEDO =1.0000   THICKNESS = 15.0000

| DEPTH | PHI | | 1 | 2 | 3 | 4 | 5 | 6 | 7 |
|---|---|---|---|---|---|---|---|---|---|
| 0. | INFINITE | B | 2.9193E 00 | 2.3071E 00 | 2.0445E 00 | 1.9030E 00 | 1.8172E 00 | 1.7647E 00 | 1.7371E 00 |
| | | H | 3.9299E 01 | 7.7379E 00 | 3.3661E 00 | 2.0000E 00 | 1.4226E 00 | 1.1486E 00 | 1.0261E 00 |
| | | | 6.5523E-02 | 7.9682E-02 | 9.9689E-02 | 1.2257E-01 | 1.4489E-01 | 1.6312E-01 | 1.7434E-01 |
| | | | 0. | 0. | 0. | 0. | 0. | 0. | 0. |
| 1.00 | 1.5880E 00 | B | 1.5836E 00 | 1.5673E 00 | 1.5438E 00 | 1.5180E 00 | 1.4936E 00 | 1.4741E 00 | 1.4622E 00 |
| | | H | 1.5926E 00 | 1.6184E 00 | 1.7253E 00 | 1.7498E 00 | 1.6607E 00 | 1.5593E 00 | 1.4948E 00 |
| | | | 1.8189E-01 | 1.9301E-01 | 2.1090E-01 | 2.3246E-01 | 2.5396E-01 | 2.7173E-01 | 2.8271E-01 |
| | | | 1.7642E-01 | 1.6512E-01 | 1.4532E-01 | 1.2216E-01 | 1.0363E-01 | 9.1612E-02 | 8.5376E-02 |
| 2.00 | 1.4527E 00 | B | 1.4490E 00 | 1.4380E 00 | 1.4194E 00 | 1.3972E 00 | 1.3752E 00 | 1.3572E 00 | 1.3461E 00 |
| | | H | 1.4557E 00 | 1.4681E 00 | 1.4929E 00 | 1.5296E 00 | 1.5392E 00 | 1.5211E 00 | 1.5011E 00 |
| | | | 2.8832E-01 | 2.9930E-01 | 3.1704E-01 | 3.3847E-01 | 3.5990E-01 | 3.7762E-01 | 3.8857E-01 |
| | | | 2.8293E-01 | 2.7193E-01 | 2.5402E-01 | 2.3199E-01 | 2.1072E-01 | 1.9464E-01 | 1.8553E-01 |
| 3.00 | 1.3422E 00 | B | 1.3395E 00 | 1.3283E 00 | 1.3104E 00 | 1.2888E 00 | 1.2673E 00 | 1.2495E 00 | 1.2385E 00 |
| | | H | 1.3449E 00 | 1.3562E 00 | 1.3748E 00 | 1.3999E 00 | 1.4216E 00 | 1.4292E 00 | 1.4283E 00 |
| | | | 3.9399E-01 | 4.0494E-01 | 4.2266E-01 | 4.4407E-01 | 4.6548E-01 | 4.8318E-01 | 4.9413E-01 |
| | | | 3.8861E-01 | 3.7766E-01 | 3.5992E-01 | 3.3837E-01 | 3.1688E-01 | 2.9963E-01 | 2.8936E-01 |
| 4.00 | 1.2356E 00 | B | 1.2329E 00 | 1.2219E 00 | 1.2041E 00 | 1.1827E 00 | 1.1613E 00 | 1.1435E 00 | 1.1326E 00 |
| | | H | 1.2383E 00 | 1.2493E 00 | 1.2672E 00 | 1.2894E 00 | 1.3115E 00 | 1.3265E 00 | 1.3330E 00 |
| | | | 4.9951E-01 | 5.1046E-01 | 5.2817E-01 | 5.4957E-01 | 5.7098E-01 | 5.8868E-01 | 5.9963E-01 |
| | | | 4.9414E-01 | 4.8319E-01 | 4.6548E-01 | 4.4405E-01 | 4.2260E-01 | 4.0501E-01 | 3.9431E-01 |
| 5.00 | 1.1298E 00 | B | 1.1271E 00 | 1.1162E 00 | 1.0985E 00 | 1.0770E 00 | 1.0556E 00 | 1.0379E 00 | 1.0270E 00 |
| | | H | 1.1325E 00 | 1.1435E 00 | 1.1612E 00 | 1.1828E 00 | 1.2045E 00 | 1.2215E 00 | 1.2309E 00 |
| | | | 6.0500E-01 | 6.1595E-01 | 6.3366E-01 | 6.5506E-01 | 6.7647E-01 | 6.9417E-01 | 7.0511E-01 |
| | | | 5.9963E-01 | 5.8869E-01 | 5.7098E-01 | 5.4957E-01 | 5.2815E-01 | 5.1047E-01 | 4.9962E-01 |
| 6.00 | 1.0243E 00 | B | 1.0216E 00 | 1.0106E 00 | 9.9293E-01 | 9.7152E-01 | 9.5011E-01 | 9.3241E-01 | 9.2146E-01 |
| | | H | 1.0270E 00 | 1.0379E 00 | 1.0556E 00 | 1.0771E 00 | 1.0986E 00 | 1.1161E 00 | 1.1265E 00 |
| | | | 7.1043E-01 | 7.2143E-01 | 7.3914E-01 | 7.6054E-01 | 7.8195E-01 | 7.9965E-01 | 8.1057E-01 |
| | | | 7.0512E-01 | 6.9417E-01 | 6.7646E-01 | 6.5506E-01 | 6.3365E-01 | 6.1595E-01 | 6.0504E-01 |
| 7.00 | 9.1877E-01 | B | 9.1608E-01 | 9.0514E-01 | 8.8743E-01 | 8.6602E-01 | 8.4462E-01 | 8.2692E-01 | 8.1597E-01 |
| | | H | 9.2145E-01 | 9.3240E-01 | 9.5011E-01 | 9.7152E-01 | 9.7152E-01 | 1.0106E 00 | 1.0214E 00 |
| | | | 8.1597E-01 | 8.2691E-01 | 8.4462E-01 | 8.6603E-01 | 8.8745E-01 | 9.0514E-01 | 9.1601E-01 |
| | | | 8.1060E-01 | 7.9965E-01 | 7.8195E-01 | 7.6054E-01 | 7.3914E-01 | 7.2143E-01 | 7.1050E-01 |
| 8.00 | 8.1328E-01 | B | 8.1060E-01 | 7.9965E-01 | 7.8195E-01 | 7.6054E-01 | 7.3914E-01 | 7.2143E-01 | 7.1050E-01 |
| | | H | 8.1597E-01 | 8.2691E-01 | 8.4462E-01 | 8.6603E-01 | 8.8745E-01 | 9.0514E-01 | 9.1601E-01 |
| | | | 9.1608E-01 | 9.3240E-01 | 9.5011E-01 | 9.7152E-01 | 9.2960E-01 | 1.0106E 00 | 1.0214E 00 |
| | | | 9.1608E-01 | 9.0514E-01 | 8.8743E-01 | 8.6602E-01 | 8.4462E-01 | 8.2692E-01 | 8.1597E-01 |
| 9.00 | 7.0780E-01 | B | 7.0512E-01 | 6.9417E-01 | 6.7646E-01 | 6.5506E-01 | 6.3365E-01 | 6.1595E-01 | 6.0504E-01 |
| | | H | 7.1049E-01 | 7.2143E-01 | 7.3914E-01 | 7.6054E-01 | 7.8195E-01 | 7.9965E-01 | 8.1057E-01 |
| | | | 1.0270E 00 | 1.0379E 00 | 1.0556E 00 | 1.0771E 00 | 1.0986E 00 | 1.1161E 00 | 1.1265E 00 |
| | | | 1.0216E 00 | 1.0106E 00 | 9.9293E-01 | 9.7152E-01 | 9.5011E-01 | 9.3241E-01 | 9.2146E-01 |
| 10.00 | 6.0232E-01 | B | 5.9963E-01 | 5.8869E-01 | 5.7098E-01 | 5.4957E-01 | 5.2815E-01 | 5.1047E-01 | 4.9962E-01 |
| | | H | 6.0500E-01 | 6.1595E-01 | 6.3366E-01 | 6.5506E-01 | 6.7647E-01 | 6.9417E-01 | 7.0511E-01 |
| | | | 1.1325E 00 | 1.1435E 00 | 1.1612E 00 | 1.1828E 00 | 1.2045E 00 | 1.2215E 00 | 1.2309E 00 |
| | | | 1.1271E 00 | 1.1162E 00 | 1.0985E 00 | 1.0770E 00 | 1.0556E 00 | 1.0379E 00 | 1.0270E 00 |
| 11.00 | 4.9683E-01 | B | 4.9414E-01 | 4.8319E-01 | 4.6548E-01 | 4.4405E-01 | 4.2260E-01 | 4.0501E-01 | 3.9431E-01 |
| | | | 4.9951E-01 | 5.1046E-01 | 5.2817E-01 | 5.4957E-01 | 5.7098E-01 | 5.8868E-01 | 5.9963E-01 |

| | | | 1.2383E 00 / 1.2329E 00 | 1.2493E 00 / 1.2219E 00 | 1.2672E 00 / 1.2041E 00 | 1.2894E 00 / 1.1827E 00 | 1.3115E 00 / 1.1613E 00 | 1.3265E 00 / 1.1435E 00 | 1.3330E 00 / 1.1326E 00 |
|---|---|---|---|---|---|---|---|---|---|
| | | X | | | | | | | |
| 12.00 | 3.9130E-01 | B | 3.8861E-01 / 3.9399E-01 | 3.7766E-01 / 4.0494E-01 | 3.5992E-01 / 4.2260E-01 | 3.3837E-01 / 4.4407E-01 | 3.1688E-01 / 4.6548E-01 | 2.9963E-01 / 4.8318E-01 | 2.8936E-01 / 4.9413E-01 |
| | | H | 1.3449E 00 / 1.3395E 00 | 1.3562E 00 / 1.3283E 00 | 1.3748E 00 / 1.3104E 00 | 1.3999E 00 / 1.2888E 00 | 1.4216E 00 / 1.2673E 00 | 1.4292E 00 / 1.2495E 00 | 1.4283E 00 / 1.2385E 00 |
| 13.00 | 2.8562E-01 | B | 2.8293E-01 / 2.6832E-01 | 2.7193E-01 / 2.9030E-01 | 2.5402E-01 / 3.1704E-01 | 2.3199E-01 / 3.3847E-01 | 2.1072E-01 / 3.5990E-01 | 1.9464E-01 / 3.7762E-01 | 1.8553E-01 / 3.8857E-01 |
| | | H | 1.4557E 00 / 1.4498E 00 | 1.4681E 00 / 1.4380E 00 | 1.4929E 00 / 1.4194E 00 | 1.5296E 00 / 1.3972E 00 | 1.5392E 00 / 1.3752E 00 | 1.5211E 00 / 1.3572E 00 | 1.5011E 00 / 1.3461E 00 |
| 14.00 | 1.7916E-01 | B | 1.7642E-01 / 1.6189E-01 | 1.6512E-01 / 1.9301E-01 | 1.4532E-01 / 2.1090E-01 | 1.2216E-01 / 2.3246E-01 | 1.0363E-01 / 2.5396E-01 | 9.1612E-02 / 2.7173E-01 | 8.5376E-02 / 2.8271E-01 |
| | | H | 1.5926E 00 / 1.5836E 00 | 1.6184E 00 / 1.5673E 00 | 1.7253E 00 / 1.5438E 00 | 1.7498E 00 / 1.5180E 00 | 1.6607E 00 / 1.4936E 00 | 1.5593E 00 / 1.4741E 00 | 1.4948E 00 / 1.4622E 00 |
| 15.00 | 6.0897E-02 | B | 0. / 6.5523E-02 | 0. / 7.9682E-02 | 0. / 9.9689E-02 | 0. / 1.2257E-01 | 0. / 1.4489E-01 | 0. / 1.6312E-01 | 0. / 1.7434E-01 |
| | | H | 3.9299E 00 / 2.9193E 00 | 7.7379E 00 / 2.3071E 00 | 3.3661E 00 / 2.0445E 00 | 2.0000E 00 / 1.9030E 00 | 1.4226E 00 / 1.8172E 00 | 1.1489E 00 / 1.7647E 00 | 1.0261E 00 / 1.7371E 00 |

ALBEDO =1.0000    THICKNESS = 20.0000

| DEPTH | PHI | | 1 | 2 | 3 | 4 | 5 | 6 | 7 |
|---|---|---|---|---|---|---|---|---|---|
| 0. | INFINITE | B | 2.9346E 00 | 2.3257E 00 | 2.0678E 00 | 1.9316E 00 | 1.8510E 00 | 1.8027E 00 | 1.7778E 00 |
| | | H | 3.9299E 01 | 2.7379E 00 | 3.3661E 00 | 2.0000E 00 | 1.4226E 00 | 1.0484E 00 | 1.0261E 00 |
| | | | 5.0228E-02 | 6.1082E-02 | 7.6420E-02 | 9.3962E-02 | 1.1107E-01 | 1.2505E-01 | 1.3364E-01 |
| | | | 0. | 0. | 0. | 0. | 0. | 0. | 0. |
| 1.00 | 1.6299E 00 | B | 1.6261E 00 | 1.6123E 00 | 1.5931E 00 | 1.5723E 00 | 1.5529E 00 | 1.5375E 00 | 1.5282E 00 |
| | | H | 1.6338E 00 | 1.6569E 00 | 1.7592E 00 | 1.7783E 00 | 1.6849E 00 | 1.5807E 00 | 1.5147E 00 |
| | | | 1.3943E-01 | 1.4796E-01 | 1.6167E-01 | 1.7820E-01 | 1.9468E-01 | 2.0830E-01 | 2.1672E-01 |
| | | | 1.3524E-01 | 1.2658E-01 | 1.1140E-01 | 9.3643E-02 | 7.9442E-02 | 7.0228E-02 | 6.5447E-02 |
| 2.00 | 1.5194E 00 | B | 1.5171E 00 | 1.5079E 00 | 1.4934E 00 | 1.4762E 00 | 1.4592E 00 | 1.4453E 00 | 1.4368E 00 |
| | | H | 1.5217E 00 | 1.5315E 00 | 1.5522E 00 | 1.5838E 00 | 1.5884E 00 | 1.5666E 00 | 1.5444E 00 |
| | | | 2.2102E-01 | 2.2943E-01 | 2.4303E-01 | 2.5947E-01 | 2.7589E-01 | 2.8947E-01 | 2.9787E-01 |
| | | | 2.1689E-01 | 2.0846E-01 | 1.9473E-01 | 1.7783E-01 | 1.6153E-01 | 1.4921E-01 | 1.4222E-01 |
| 3.00 | 1.4335E 00 | B | 1.4314E 00 | 1.4229E 00 | 1.4091E 00 | 1.3925E 00 | 1.3760E 00 | 1.3623E 00 | 1.3539E 00 |
| | | H | 1.4357E 00 | 1.4443E 00 | 1.4588E 00 | 1.4789E 00 | 1.4955E 00 | 1.4991E 00 | 1.4958E 00 |
| | | | 3.0202E-01 | 3.1042E-01 | 3.2400E-01 | 3.4041E-01 | 3.5682E-01 | 3.7040E-01 | 3.7879E-01 |
| | | | 2.9790E-01 | 2.8950E-01 | 2.7590E-01 | 2.5939E-01 | 2.4292E-01 | 2.2969E-01 | 2.2182E-01 |
| 4.00 | 1.3516E 00 | B | 1.3495E 00 | 1.3411E 00 | 1.3274E 00 | 1.3110E 00 | 1.2945E 00 | 1.2809E 00 | 1.2725E 00 |
| | | H | 1.3536E 00 | 1.3621E 00 | 1.3759E 00 | 1.3930E 00 | 1.4101E 00 | 1.4210E 00 | 1.4251E 00 |
| | | | 3.8291E-01 | 3.9131E-01 | 4.0488E-01 | 4.2129E-01 | 4.3770E-01 | 4.5127E-01 | 4.5966E-01 |
| | | | 3.7880E-01 | 3.7040E-01 | 3.5683E-01 | 3.4040E-01 | 3.2395E-01 | 3.1047E-01 | 3.0227E-01 |
| 5.00 | 1.2704E 00 | B | 1.2684E 00 | 1.2600E 00 | 1.2464E 00 | 1.2299E 00 | 1.2135E 00 | 1.2000E 00 | 1.1916E 00 |
| | | H | 1.2725E 00 | 1.2809E 00 | 1.2945E 00 | 1.3111E 00 | 1.3278E 00 | 1.3407E 00 | 1.3475E 00 |
| | | | 4.6378E-01 | 4.7217E-01 | 4.8575E-01 | 5.0215E-01 | 5.1856E-01 | 5.3213E-01 | 5.4053E-01 |
| | | | 4.5967E-01 | 4.5127E-01 | 4.3770E-01 | 4.2129E-01 | 4.0487E-01 | 3.9132E-01 | 3.8300E-01 |
| 6.00 | 1.1895E 00 | B | 1.1874E 00 | 1.1790E 00 | 1.1654E 00 | 1.1490E 00 | 1.1326E 00 | 1.1191E 00 | 1.1107E 00 |
| | | H | 1.1915E 00 | 1.1999E 00 | 1.2135E 00 | 1.2300E 00 | 1.2465E 00 | 1.2599E 00 | 1.2477E 00 |
| | | | 5.4464E-01 | 5.5303E-01 | 5.6660E-01 | 5.8301E-01 | 5.9942E-01 | 6.1299E-01 | 6.2138E-01 |
| | | | 5.4053E-01 | 5.3213E-01 | 5.1856E-01 | 5.0215E-01 | 4.8574E-01 | 4.7217E-01 | 4.6381E-01 |
| 7.00 | 1.1086E 00 | B | 1.1065E 00 | 1.0981E 00 | 1.0846E 00 | 1.0682E 00 | 1.0518E 00 | 1.0382E 00 | 1.0298E 00 |
| | | H | 1.1107E 00 | 1.1190E 00 | 1.1326E 00 | 1.1490E 00 | 1.1655E 00 | 1.1790E 00 | 1.1872E 00 |
| | | | 6.2550E-01 | 6.3389E-01 | 6.4746E-01 | 6.6387E-01 | 6.8028E-01 | 6.9385E-01 | 7.0224E-01 |
| | | | 6.2138E-01 | 6.1299E-01 | 5.9942E-01 | 5.8301E-01 | 5.6660E-01 | 5.5303E-01 | 5.4465E-01 |
| 8.00 | 1.0277E 00 | B | 1.0257E 00 | 1.0173E 00 | 1.0037E 00 | 9.8730E-01 | 9.7089E-01 | 9.5732E-01 | 9.4893E-01 |
| | | H | 1.0298E 00 | 1.0382E 00 | 1.0518E 00 | 1.0682E 00 | 1.0846E 00 | 1.0981E 00 | 1.1065E 00 |
| | | | 7.0636E-01 | 7.1475E-01 | 7.2832E-01 | 7.4473E-01 | 7.6114E-01 | 7.7471E-01 | 7.8310E-01 |
| | | | 7.0224E-01 | 6.9385E-01 | 6.8028E-01 | 6.6387E-01 | 6.4746E-01 | 6.3389E-01 | 6.2550E-01 |
| 9.00 | 9.4687E-01 | B | 9.4482E-01 | 9.3642E-01 | 9.2285E-01 | 9.0646E-01 | 8.9004E-01 | 8.7646E-01 | 8.6807E-01 |
| | | H | 9.4893E-01 | 9.5732E-01 | 9.7089E-01 | 9.8730E-01 | 1.0037E 00 | 1.0173E 00 | 1.0257E 00 |
| | | | 7.8721E-01 | 7.9561E-01 | 8.0918E-01 | 8.2559E-01 | 8.4199E-01 | 8.5556E-01 | 8.6395E-01 |
| | | | 7.8310E-01 | 7.7471E-01 | 7.6114E-01 | 7.4473E-01 | 7.2832E-01 | 7.1475E-01 | 7.0636E-01 |
| 10.00 | 8.6601E-01 | B | 8.6396E-01 | 8.5556E-01 | 8.4199E-01 | 8.2559E-01 | 8.0918E-01 | 7.9561E-01 | 7.8721E-01 |
| | | H | 8.6807E-01 | 8.7646E-01 | 8.9004E-01 | 9.0644E-01 | 9.2285E-01 | 9.3642E-01 | 9.4481E-01 |
| | | | 8.6807E-01 | 8.7646E-01 | 8.9004E-01 | 9.0644E-01 | 9.2285E-01 | 9.3642E-01 | 9.4481E-01 |
| | | | 8.6396E-01 | 8.5556E-01 | 8.4199E-01 | 8.2559E-01 | 8.0918E-01 | 7.9561E-01 | 7.8721E-01 |
| 11.00 | 7.8516E-01 | B | 7.8310E-01 | 7.7471E-01 | 7.6114E-01 | 7.4473E-01 | 7.2832E-01 | 7.1475E-01 | 7.0636E-01 |
| | | H | 7.8721E-01 | 7.9561E-01 | 8.0918E-01 | 8.2559E-01 | 8.4199E-01 | 8.5556E-01 | 8.6395E-01 |

*Dense numerical appendix table (scientific notation). Each x‑value has B and H rows, each printed as two lines spanning seven data columns. The two top lines (marked I) continue the x = 11 entry from the preceding page.*

```
  x        value     row    C1          C2          C3          C4          C5          C6          C7

 (I)                        9.4893E-01  9.5732E-01  9.7089E-01  9.8730E-01  1.0037E 00  1.0173E 00  1.0257E 00
                            9.4482E-01  9.3642E-01  9.2285E-01  9.0644E-01  8.9004E-01  8.7646E-01  8.6807E-01

12.00    7.0430E-01   B     7.0224E-01  6.9385E-01  6.8028E-01  6.6387E-01  6.4746E-01  6.3389E-01  6.2550E-01
                            7.0636E-01  7.1475E-01  7.2832E-01  7.4473E-01  7.6114E-01  7.7471E-01  7.8310E-01
                      H     1.0298E 00  1.0382E 00  1.0518E 00  1.0682E 00  1.0846E 00  1.0981E 00  1.1065E 00
                            1.0257E 00  1.0173E 00  1.0037E 00  9.8730E-01  9.7089E-01  9.5732E-01  9.4893E-01

13.00    6.2344E-01   B     6.2138E-01  6.1299E-01  5.9942E-01  5.8301E-01  5.6660E-01  5.5303E-01  5.4465E-01
                            6.2550E-01  6.3389E-01  6.4746E-01  6.6387E-01  6.8028E-01  6.9385E-01  7.0224E-01
                      H     1.1107E 00  1.1190E 00  1.1326E 00  1.1490E 00  1.1655E 00  1.1790E 00  1.1872E 00
                            1.1065E 00  1.0981E 00  1.0846E 00  1.0682E 00  1.0518E 00  1.0382E 00  1.0298E 00

14.00    5.4258E-01   B     5.4053E-01  5.3213E-01  5.1856E-01  5.0215E-01  4.8574E-01  4.7217E-01  4.6381E-01
                            5.4464E-01  5.5303E-01  5.6660E-01  5.8301E-01  5.9942E-01  6.1299E-01  6.2138E-01
                      H     1.1916E 00  1.1999E 00  1.2135E 00  1.2300E 00  1.2465E 00  1.2599E 00  1.2677E 00
                            1.1874E 00  1.1790E 00  1.1655E 00  1.1490E 00  1.1326E 00  1.1191E 00  1.1107E 00

15.00    4.6172E-01   B     4.5967E-01  4.5127E-01  4.3770E-01  4.2129E-01  4.0488E-01  3.9131E-01  3.8300E-01
                            4.6378E-01  4.7217E-01  4.8575E-01  5.0215E-01  5.1856E-01  5.3213E-01  5.4053E-01
                      H     1.2725E 00  1.2809E 00  1.2945E 00  1.3110E 00  1.3278E 00  1.3407E 00  1.3475E 00
                            1.2684E 00  1.2600E 00  1.2465E 00  1.2299E 00  1.2135E 00  1.1999E 00  1.1916E 00

16.00    3.8086E-01   B     3.7880E-01  3.7040E-01  3.5683E-01  3.4041E-01  3.2395E-01  3.1047E-01  3.0227E-01
                            3.8291E-01  3.9131E-01  4.0408E-01  4.2129E-01  4.3770E-01  4.5127E-01  4.5966E-01
                      H     1.3536E 00  1.3621E 00  1.3759E 00  1.3930E 00  1.4101E 00  1.4210E 00  1.4251E 00
                            1.3495E 00  1.3411E 00  1.3110E 00  1.2999E 00  1.2845E 00  1.2809E 00  1.2725E 00

17.00    2.9996E-01   B     2.9790E-01  2.8950E-01  2.7594E-01  2.5939E-01  2.4292E-01  2.2969E-01  2.2182E-01
                            3.0202E-01  3.1042E-01  3.2400E-01  3.4041E-01  3.5682E-01  3.7040E-01  3.7879E-01
                      H     1.4314E 00  1.4443E 00  1.4588E 00  1.4780E 00  1.4955E 00  1.4991E 00  1.4958E 00
                            1.4357E 00  1.4229E 00  1.4091E 00  1.3925E 00  1.3760E 00  1.3623E 00  1.3539E 00

18.00    2.1895E-01   B     2.1689E-01  2.0866E-01  1.9473E-01  1.7783E-01  1.6153E-01  1.4921E-01  1.4222E-01
                            2.2102E-01  2.2943E-01  2.4303E-01  2.5947E-01  2.7589E-01  2.8947E-01  2.9787E-01
                      H     1.5217E 00  1.5315E 00  1.5522E 00  1.5838E 00  1.5884E 00  1.5666E 00  1.5446E 00
                            1.5171E 00  1.5079E 00  1.4934E 00  1.4762E 00  1.4592E 00  1.4453E 00  1.4368E 00

19.00    1.3734E-01   B     1.3524E-01  1.2658E-01  1.1140E-01  9.3643E-02  7.9442E-02  7.0228E-02  6.5447E-02
                            1.3943E-01  1.4796E-01  1.6167E-01  1.7820E-01  1.9468E-01  2.0830E-01  2.1672E-01
                      H     1.6338E 00  1.6569E 00  1.7592E 00  1.7783E 00  1.6849E 00  1.5807E 00  1.5147E 00
                            1.6261E 00  1.6123E 00  1.5931E 00  1.5723E 00  1.5529E 00  1.5375E 00  1.5282E 00

20.00    4.6683E-02   B     5.0228E-02  6.1082E-02  0.          0.          1.1107E-01  1.2505E-01  1.3364E-01
                            3.9299E-01  7.7379E-01  7.6420E-01  2.0000E 00  1.4226E 00  1.1484E 00  1.0261E 00
                      H     2.9346E 00  2.3257E 00  2.0678E 00  1.9316E 00  1.8510E 00  1.8027E 00  1.7778E 00
```

INDEX

# 7 DAY USE
## RETURN TO
# ASTRON-MATH-STAT. LIBRARY
### Tel. No. 642-3381

This publication is due before Library closes on
the LAST DATE and HOUR stamped below.

| | |
|---|---|
| JAN 27 1978 | |
| Feb 3 78 | |
| FEB 9 1979 | |
| APR 28 1979 | |
| Due end of SUMMER quarter | |
| Subject to recall after— | |
| JUL 3 1 1984 | |
| NOV 28 1986 | |
| | |
| | |
| | |
| | |
| | |
| | |
| | |

RB17-5m-2'75
(S4013s10)4187—A-32

General Library
University of California
Berkeley